George Sinclair

The Hydrostaticks

or, the Weight, Force and Pressure of fluid Bodies

George Sinclair

The Hydrostaticks
or, the Weight, Force and Pressure of fluid Bodies

ISBN/EAN: 9783337345778

Printed in Europe, USA, Canada, Australia, Japan

Cover: Foto ©berggeist007 / pixelio.de

More available books at **www.hansebooks.com**

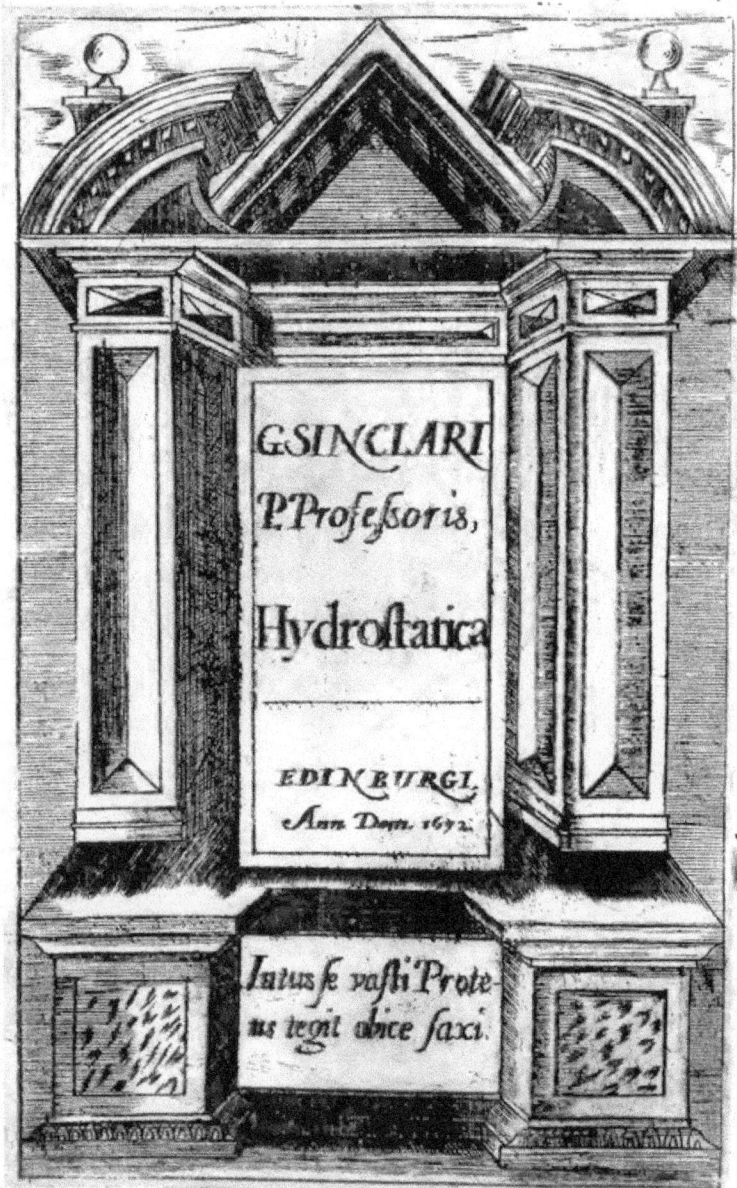

G.SINCLARI

P. Professoris,

Hydroſtatica

EDINBURGI,
Ann. Dom. 1672.

Intus se vaſti Prote-
us tegit alice ſaxi.

THE
HYDROSTATICKS;
OR,
The *Weight*, *Force*, and *Pressure* of
FLUID BODIES,
Made evident by *Physical*, and
Sensible Experiments.

TOGETHER
VVith some *Miscellany Observati-*
ons, the last whereof is a short
History of *Coal*, and of all the
Common, and *Proper Accidents* thereof;
a Subject never treated of before.

By *G. S.*

EDINBURGH,
Printed by *George Swintoun*, *James Glen*, and
Thomas Brown: Anno DOM. 1672.

SPES DURA

To my very Honourable, and Noble LORD,

ROBERT
VISCOUNT of OXFUIRD,
LORD MACKGILL of COUSLAND, &c.

My Noble Lord,

THe firſt application I make, is for *pardon*, that I have adventured to prefix your *name* to the *Frontiſpice* of this Work, which in it ſelf, cannot be thought worthy of your *Truſt* and *Protection* ; there being no proportion between the greatneſs of your *Merit*, and ſo mean an *Oblation*; ſave what flows from the Nobleneſs of the *Subject*, and the ſincerity of his reſpects who preſents it. It is truly a part of *Philoſophy*, that was

2   never

never much Cultivated, but of late, except in a more abſtract and ſubtil way, which did render it leſs uſeful; but is now more improven by ſenſible *Manifeſtations* of the Soveraign Miſtriſs of Arts, NATURE her ſelf. There are indeed (my Lord) many excellent *Sciences,* which do merit the favour of your Lordſhips ſtudies, and by which your Noble Accompliſhments might be more improven; yet I am bold to affirm, you cannot apply your *Noble Mind* to any part of *Philoſophy,* where you will find more Pleaſure, with leſs Pains; more evidence of Reaſon, with leſs Difficulty.

The famous *Gregorio Leti,* was ſo much an admirer of your *Vertues,* that he ſheltered under your *Patrociny,* his *Vita Di Siſto quinto Pontefece Romano.* And if you were able to protect an envyed *Italian* in *Italy,* much more may I expect full ſecurity from your Name in *Scotland,* where your intereſt and relations are ſo conſiderable. And if he, who only look'd upon your *Vertuous Mind,* while it was but bloſſoming, was ſo much perſwaded to judge none

more

more fit to Receive, Protect, and Claim his *Labours*, much more *I*, who have feen the accomplifhment of your *Vertues at home*. *I* have likewife very much confidence of your Noble and Candid Difpofition to admit this into your *Favour*, and affurance of your Affection and Skill, to Love it, and Underftand it; both which are confpicuous, the firft in your encouragement to all *Learning*, the other in your Capacity and Underftanding to comprehend, whatever you encourage.

Though (my Lord) I have been much emboldened to offer this *Dedication* to your Lordfhip, upon the account of your own *Heroick Vertues*, yet *I* muft not pafs over in filence, a moft fpecial Motive, which to me fhall be the laft, fparing to exprefs all the great Caufes oblieging me fo to do, and that is, the Memory of your VVorthy and neareft *Relations*, who are, *my Lord your Father*, *Grandfather*, and *Great-Grand-father*, not only memorable for their Vertue and Learning, and peculiar Endowments, whereby they were thought

thought worthy to ſerve their *King* and *Countrey*, in *Council*, and *Honourable Courts of Juſtice*
for theſe many years, but for the *Dignity*, and
Antiquity of their famous *Anceſtours*. How old
your Lordſhips Name is, *Buchanan* teſtifies in
the cloſe of the *Second Book* of his *Hiſtory*, who
writeth thus, *Certè Gildus vetus eſt in Scotia No-
men, ut vetus Mackgildorum, ſive Mackgillorum
gens indicat: è cujus poſteris honeſtæ adhuc in Scotia & Anglia ſunt familiæ.* That is, *Surely*
Gild *is an ancient Name in* Scotland, *as witneſs
the old Family of* Mackgilds, *or* Mackgills:
of whoſe Poſterity there are yet in Scotland *and*
England *many Families of good account.* And
as an inſtance of this, the ſame Author tells
us of the Great *Thane of* Galloway, *Mackgil-
lum Gallovidiæ longè Potentiſsimum*, in the life of
Mackbeth, who by this *Uſurper* was put to
death for his adherence to his *Prince*, from
whom your Lordſhip, and your worthy *Pro-
genitors* are Lineally deſcended, and of whom
Buchanan meant in the foregoing paſſage, ſince
our Predeceſſors flouriſht in his time; your
Great-

Great-Grand-Father having then been His *Majesties Advocat,* his Brother *Lord Register.*

Having now (my Noble Lord) laid before you so many considerable Motives, which I humbly desire may prevail, I cannot but make my next *Application* for *Acceptance,* and seriously intreat this *Work* may be received into the Tuition of your *Favour,* and get a full *Protection* from the *Censorious,* and being enlightned with the splendor of your *Name,* and receiving the impression of your *Authority* upon it, may safely pass thorow the VVorld, for which singular Favour, I shall fervently wish to your Self and Noble *Family,* all Prosperity, and Happiness, and shall think my self very happy under the Character of,

My Noble Lord,

Edinburgh, May 20. (*the day of your Lo. Birth and Majority*) 1672.

Your Lo. most humble and much oblieged Servant,

GEORGE SINCLAR.

TO

TO THE
READER.

Courteous Reader,

I Shall not detain thee in the entry with a long Preface, *but give a short account of what is needful to be known, of the* Cause, Occasion, *and* Matter *of the following* Treatise. *After the publication of my last* Piece, about the Weight and Pressure of the Air, *I found it needful to treat of the* Pressure of the Water, *because of the near relation between the two : the operations, and effects of both depending almost upon the same* Principles and Causes. *And that there are many things, which cannot throughly be understood, of the* Pressure *of the* Air, *without the knowledge of the* Pressure *of the* Water : *therefore to make the first the more evident, I have spoken of the second : the effects and operations of* Hydrostatical Experiments, *being more conspicuous and sensible, then the effects and operations of the other.*

The Occasion *was some spare time I had now and then, for making some* Trials : *part whereof are published here ; the rest being rather some productions of* Reason, *attentively exercised on that* Subject ; *which notwithstanding may be called* Experiments, *though never actually tried, nor haply can be, because of some accidental impediments : yet supposing they were, I make it evident, that such and such* Phenomena *would follow, whence many necessary conclusions are inferred.*

As for the subject matter, there are first, moe then thirty Theorems *in order to the* Pressure *of* Fluid Bodies, *as* Air, Water, *and* Mercury, *which in effect are nothing else, but so many conclusions rationally deduced from various and diverse effects of* Aerostatical, Hydrostatical, *and* Hydrargyrostatical Experiments, *which for the most part, I have tried my self.*

There are next twenty Experiments *briefly described, by their own* distinct

¶ ¶

To the Reader.

diſtinct Schematiſms : *their* Phenomena, *according to the Laws of the* Hydroſtaticks *are* ſalved, *and ſeveral new concluſions inferred. A* Propoſal *is likewiſe made of a more convenient* Engine *for* Diving. *Here, ſeveral difficulties are propoſed, and anſwered, and all the obvious* Phenomena *of* Diving *explicated. If the* Lead *which ſinks the* Ark, *be judged too weighty, and big, which may render it not ſo tractable, and likewiſe hinder the* Ark *from going ſo near to the ground, as is deſirable, and in ſome meaſure ſtop the ſight, (which troubles are (I ſuppoſe) incident to the* Bell *alſo) it may be reduced to a far leſs weight, and quantity, and the overplus being made ſquare and thin pieces, may like the mouth of the* Ark *without, between* P Q *and* L M, *according to the Figure* 25, *or may be put to, or taken away at pleaſure. The* Bell *may have likewiſe in ſtead of this troubleſome* Foot-board, *a weighty* Ring *of* Lead, *or two, to go round about the orifice without, in form of a* Girth, *or* Belt, *which may ſlip off and on at pleaſure, and will as conveniently ſink it, as if it had a weight appended: the* Foot-board *then may be of any form, quantity, or weight you pleaſe.*

There are thirdly ſome *Miſcellany* Obſervations, *the deſign of which is only* Philoſophical. *Some of them are* Experiments *made with the* Air-pump, *which I have adventured to inſert here, even though the Noble* Mr. Boyl *hath given an account of many. But becauſe the* Engine *was offered to me by the* Laird *of* Salton, *a Gentleman of a choiſe Spirit, I could not, but in obedience to his commands make uſe of it, and ſhew him the Product. There are alſo two or three* Obſervations *in the cloſe, as that of the* Primum vivens *in* Animals : *of the* Aliment, *and growth of plants: and of the motion of the* Aliment *in* Trees. *To all which is added a ſhort* Hiſtory *of* Coal, *which I hope will be acceptable to ſome ; this ſo needful a ſubject, never being treated of before by any. In it, mention is made of things common to* Coal *in general, as* Dipps, Riſings, *and* Streeks. Next, *of* Gaes, *or* Dykes, *which prove ſo troubleſome ſometimes to the working of* Coal. Thirdly, *of* Damps, *and* Wild-fire. Next, *a method is taught for trying of* Grounds, *where never any* Coal *was diſcovered before. And laſtly, the manner how* Levels, *or* Conduits *under-ground, ought to be carried on, for draining the* Coal, *and freeing it of* Water.

When

To the Reader.

When this Book was first committed to the Press, *I sent an* intimation *thereof to some of my friends, for their encouragment to it, a* Practice *now common, and commendable, which hath not wanted a considerable success, as witness the respect of many worthy persons, to whom I am obliged. But there is a* Generation, *that rather, than they will encourage any new* Invention, *set themselves by all means to detract from it, and the* Authors *of it: so grieved are they, that ought of this kind should fall into the hands of any, but their own. And therefore, if the* Author *shall give but the title of* New *to his Invention, though never so deservedly, they fly presently in his throat, like so many* Wild-Catts, *studying either to* Ridicule *his work altogether, a trade that usually, the person of weakest abilities, and most empty heads, are better at, than learned men, like those Schollars, who being nimble in putting tricks, and impostures upon their Condisciples, were dolts, as to their Lesson, or else fall upon it with such snarling, and carping, as discover, neither ingenuity, nor ingeniousness, but a sore sickness, called* Envy.

In the Intimation, *I affirmed, that the Doctrine concerning the* Weight, *and Pressure of the* Water *was* New. *This one word, like a spark of Fire falling accidentally among* Powder, *hath been the occasion of so much debate. Their ground is, because they look upon the* Hydrostaticks, *as a* Science *long ago perfected, seing* Archimedes *2000 years ago hath demonstrat the* Water *to have a* Pressure, *and some others since, as* Stevinus. *They affirm likewise, that all the* Theorems, *and* Experiments, *that are here, are either deduceable from* Archimedes, *and* Stevinus, *or are the same with theirs. If these Gentlemen had suspended their judgment, till this Book had been published, I suspect they would not have spoken so confidently. For* Archimedes *his propositions, they are but few, and proven (as* Mr Boyl *saith) by no very easie demonstrations, which have more of Geometrical subtility, than usefulness in them. But these, which are here proposed, are not only useful, but evidently evicted by reason, and sensible* Experiments, *even to the meanest capacities. And though some of mine, may (perhaps) co-incide with some of his, which to me is but accidental, yet our way of procedour is* toto Cœlo *different. His way is more* Speculative: *this is more* Practical. *His demonstrations*

¶¶ 2

are

To the Reader.

are Geometrical *: these are* Physical. *His propositions are but for the use of a few: these are for the* use *of all. His are not illustrated, and confirmed by* Hydrostatical Experiments *: these are.*

Stevinus *a late Writer keeps that same method. Yet I judge it easie to let see, even in the entry, how little cogent some of his demonstrations are, without derogating from such a Learned Man. He hath indeed some* Pragmatical Examples (*as he calls them*) *for illustrating some of his* Geometrical Propositions , *anent the* Pressure *of the* Water *; but I leave them to be considered by the judicious and understanding. Again, in this* Method *, I am yet as much different from others, who have written lately, as from these I have been speaking of. For, I not only treat of the* Pressure *of the* Water *, but takes in with it , the* Pressure *of the* Air *joyntly ; since to explicat sufficiently the* Phenomena *of the* Hydrostaticks *, without it , it is impossible. And yet farder, I not only counterpoise* Air *with* Water *, but* Air *with* Mercury *, and* Water *with* Mercury *, by which means several mysteries, and secrets in this Art, are discovered.*

There are several Inventions found out of late in the Hydrostaticks *, whose events and effects, cannot be clearly deduced from the grounds of* Archimedes, *and* Stevinus, *who had not that clear discovery (for ought we know) of the* Pressure of the Air, *that some now have, without which, these effects can never be sufficiently explained. And who doubts, but others afterwards, may make farder discoveries, and profit the world yet more, with their Inventions, then any have yet done. Is then the* Hydrostaticks, *a* Science long ago perfected ? To this Pedantick Conceit, *I must again oppose the judgment of* Mr. Boyl, *who saith moreover, that* the usefulness of this part of Philosophy hath been scarce known any farder than by name, even to the generality of learned men.

But let us suppose, that the notion *of the* Pressure *of the* Water, *is of an old date, even as old as the* Flood *(for* Noah *surely knew, that the* Pressure of the Water, *would sustain the* Ark *) and (giving, but not granting) that* Archimedes 2000 *years ago hath written all the* Principles of the Hydrostaticks, *doth this hinder any man now, from deducing* new Conclusions *from these* old Principles ? *But there is here, no such thing: for neither in this, nor in my last* Piece, *are my*

Adver-

To the Reader.

Adverfaries *able to* trace me. *'Tis like the purpoſes would have been ſo much the better, if I had followed other mens foot ſteps: and it is like they might have been ſo much the worſe. I doubt not, but I have lighted upon other mens thoughts in ſome things: and others writing on this ſame ſubject, who perhaps are my* Antipodes *, may fall upon mine.* My Antagoniſts *affirm, they are able to deduce all my* Theorems, *and the events of all my* Experiments *from the grounds of* Archimedes *and* Stevinus. *If* they take not their word again, *I hope they will do it; for now I put them to it. And though they ſhould, (which I am not affraid they ſhall do in haſte) yet they muſt prove next, that theſe* Theorems *and* Concluſions, *ſo deduced, are not new, which all their* Logick *will not prove. But what if we do more, (ſay they) even overthrow many of all your* Aeroſtatical *and* Hydroſtatical Experiments, *in this, and in your laſt* Peice? *I give you liberty, and for your hire, a* Guiny *for each* Theorem, *or* Experiment, *you are able to ranſack, in either of the two* Books, *though they come near to an hundred. But, ye muſt oblige your ſelves (my* Maſters) *to do it with* Reaſon, *laying aſide your* Sophiſtry *and* Canina eloquentia. *And this I offer, Reader, that I may reduce them, to a better* humour, *and encourage them to leave off flyting, and only uſe reaſon. Neither muſt they be like the* Waſp, *that only lights upon the* ſore place. *But if they love to kindle any more fire, they will find me proof againſt it. If it burn them, it ſhall not heat me. Neverthelſſ, if they love to* juik *under deck, like* Green-horns, *having no courage in themſelves, or confidence in their cauſe, they muſt excuſe me, if at laſt, I write their names upon a* Ticket, *and bring them above deck. This is all I have to ſay, at preſent (* Reader *) and I bid thee farewell.*

E R R A T A.

Pag. 22. lin. 8. *for* weight *read* benſil. Pag. 185. lin. 24. *for* E H. *read* F H. Pag. 235. lin. 24. *for* 500. *read* 5000. Pag. 307. lin. 26. read promoting. Pag. 313. lin. 22. *read* reflection. *Ibid.* lin. 25. *read* elaborarint. Pag. 317. lin. 2. *read* & magna.

Note, that in placing the Figures, the 12, that ſhould have the fourth place in the third Plate, hath the firſt place in the fourth.

Contents

Contents of the Experiments.

Contents of the MISCELLANY OBSERVATIONS.

In Auctorem & Opus Encomiasticon.

ÆTheris expanſi, vitrei Maris Antitalanton,
Peroledos, Elaſin, *Fluidarum ritè videntes*,
Ingenio patefaôa tuo, Magnalia rerum,
Laudarunt alacres Galli , Belgæque *ſagaces.*
Aggrederis nunc Arte Novâ, trutinare profundi
Corpora, ſubmerſas quondam producere Gazas,
Tollere demerſis ingentia pondera Cupis.
Gas *fracidum in* Cryptis *ortum Foſſoribus atrox*,
Submiſſo in Fundos Auræ renovante Flabello,
Propulſare doces, Lithanthracumque *Cavernæ*
Quêis foveantur Aquis, quo tendant, unde oriantur,
Ordine quò circum Saxorum ſtrata recumbant.
Quòdbenè cœpiſti Naturæ cunôa foventis
Munera ſolerti perge Illuſtrare Matheſi.

GEORGIUS HEPBURNUS, *M. D.*
à Monachagro.

READER,

[ ]

To the Reader.

Reader,

THat thou mayeſt know, by one word more, how uſeful this part of *Philoſophy* is, and how far from being
a *Science long ago perfected*, take but this following *propoſal*, lately, ſince my Book came to a cloſe, communicated
to me by a *Friend*, which, by his allowance, I have publiſhed, reſerving the Anſwer to himſelt, the Author thereof.

Brother,

BY *what you have publiſhed in your* Ars Nova & Magns,
and this Book, I *have been led to this* Invention, *to
beget within the Bowels of the Sea, a Power, or Force, which
with great ſafety, and eaſe, ſhall bring up the greateſt weight,
that can be ſunk therein:* ad data quæcunque pondera de-
merſa, in Maris viſceribus Potentiam producere, quæ mo-
do ſecuro, & facili, è tundo cujuſvis altitudinis ad ſum-
mum, ipſa evehat. I *drew a Letter one night, ſhewing the
way how this might be done, which* I *communicated to you,
that it might have been Printed with your Book: but after
ſecond thoughts,* I *judged it more meet to keep it up for a
time, and that it ſhould be ſet forth by way of* Proposal *only
at the firſt, by*

Ormiſton,   Your Brother,
May 20. 1672.   Mr. *John Sinclar*.

This *New Invention*, though Hydroſtatical, is truly
Mechanical, there being here a *Tondus* and a *Potentia*, whoſe
operations depends upon Mechanical Principles. But in

¶ ¶ ¶   ſeveral

several respects it is far more admirable, than the most part of the Mechanical Engines, which are look'd upon as stupendious. Many things, almost incredible, are reported of *Archimedes*, which he admirably brought about, by his Mechanical *Powers*; but I am confident, that by this *Invention*, as great a weight may be lifted, if not greater, as the *Power* of any Mechanical Faculty can be able to move. I know, the greatest conceivable weight, may be demonstrat, to be moved by the least conceivable Power, as the *Earth*, by the *force of a mans hand*. But how is it possible to contrive Artificially, an Engine for that purpose, which will do that by *Art*, which the demonstration makes evident by reason? It was thought a great enterprize, when *Pope Sixtus* the fifth, transported an *Obelisk*, which had been long since dedicated to the memory of *Julius Cesar*, from the left side of the *Vatican*, to a more eminent place, 100 foot distant; but to raise a Ship of 1000 Tun intirely, nay, a weight 100 times greater, is surely a far greater enterprize. This *Invention* is so much the more admirable, that not only by it, any supposed weight may be lifted, but from any deepness. Though this (perhaps) cannot be done Mechanically, because of some *Physical*, or *Moral* impediment, yet according to the *Laws* of the *Hydrostaticks* it can be demonstrat, and made evident by reason. And if this be, then surely, when the *Weight* is determinat, as the burdens of all *Ships* are, and the deepness known to be within so many fathoms, this *Invention* cannot but be successful.

Though the *strength* of *Mechanical Inventions*, may be multiplied, beyond the bounds of our *Imagination*, whereby the greatest *Weight*, may be moved, by the least *Power*; yet the *Wisdom* of *God*, hath thought it fit, so to confine that knowledge, that it cannot teach, how both
of

of them, can move with the same *quickness* and *speed*. For, if that were, the very works of Nature might be overturned. Therefore, it is observable, that when a great *Weight* is moved by a small *Power*, the motion of the one, is as much flower than the motion of the other, as the *Weight* of the one, exceeds the *Force* of the other. If it were possible *Mechanically* to move the *Earth* with the *Force* of a mans hand, the motion thereof would be as much flower, than the motion of the hand, as the *Weight* of the one, exceeds the *Force* of the other, which is a great disadvantage. And as the *Weight* and *Power* do thus differ, as to *swiftness*, and *flowness* in motion, so also, as to *Space*. For, by how much the *Power* is in it self less, than the *Weight*, by so much will the bounds or *Space*, the *Weight* moves thorow, be less than the *Space*, the *Power* goes thorow. If it were possible (keeping the same instance) to move the *Earth* with a mans hand, the *Space* thorow which it passeth, would differ as much from the *Space* the hand goes thorow, as the one exceeds the other ; which is another disadvantage.

It may be thought, that if this *Invention* depend upon *Mechanical Principles*, it may be obnoxious to these abatements. I answer, though there be in it a *Pondus*, and a *Potentia*, a *Weight*, and a *Power*, this moving the other, yet it will evidently appear from Experience, that the motion of the one, is as swift as the motion of the other, and that the one moves as much *Space* and bounds in the same time , as the other, which is a great advantage. In this, it excells all the *Mechanical Powers*, and *Faculties*, that have ever yet been invented and practised. If any think, that such a device cannot be effectuat, without a considerable expence. I answer, the expence is so small, that I am ashamed to mention it. The method and manner of doing this, is most easie likewise. Neither ought this to be a ground, why any man
should

[ ]

should contemn it ; since the most useful Inventions ordinarily are performed with the greatest facility.

As it commends this part of *Philosophy* to all ingenious
Spirits, as most pleasant, and most profitable ; so it gives
a check to the *ignorant*, who look upon it as *a Science long
ago perfected.*

In praise of the AUTHOR, and his WORK.

1.

Whilst *Infant-Art no further did pretend
Then to flat notions, and a bare desire ;
What by small toyl we now do comprehend,
Our Predecessors only did admire.*

2.

*Now fruitful Reason, arm'd with powerful Art,
Uncovers Nature to each knowing eye :
Our* Author *to the World doth here impart
What was before esteem'd a mystery.*

3.

*The various motions of that Element,
Whose liquid form gives birth to much debate ;
By demonstration he doth represent,
Unfolding th'intrigues of that subtil state.*

4.

*The Water's Course, and Sourse, from whence they flow,
By him to th'sense so clearly are display'd :
Their current Weight, and Measure now we know,
'Tis no more secret, but an open Trade.*

W. C.

HYDRO-

Hydroſtatical
THEOREMS,

Containing ſome uſeful Principles in order to that excellent Doctrine, anent
the wonderful *Weight*, *Force*, and *Preſſure* of
the Water in its own Element.

THEOREM I.

*In all Fluids, beſides the firſt and viſible Horizontal ſur-
face, there are many moe* imaginary, *yet* real.

Figure 1.

OR the better underſtanding the
following Experiments, it is
needful to premit the ſubſequent Theorems; the firſt whereof is, that in all Fluid bodies,
ſuch as Air, Water, and Mercury,
or any other liquid, there is beſides the firſt and viſible ſurface,
innumerable moe imaginary, under that firſt, yet real, as may
be ſeen from the following Schematiſm, which repréſents
a Veſſel full of Water, where beſides the firſt ſurface

A   A B

A B C D, there is a ſecond E F G H, and a third I K L M, and ſo downward, till you come to the bottom. This holds true, not only in Water, but in Air alſo, or in any other Fluid body whatſoever. I call the under-ſurfaces imaginary, not becauſe they are not real; for true and real effects are performed by them; but becauſe they are not actually diſtinguiſhed amongſt themſelves, but only by the Intellect.

THEOREM II.

In all Fluids, as it is needful to conceive Horizontal Plains, ſo it is needful to conceive Perpendicular Pillars, cutting theſe Plains at right Angles.
Figure 1.

THis Propoſition is likewiſe needful for underſtanding the following Doctrine, anent the Preſſure of the Water: for in it, as in all Fluids, though there be not Columes or Pillars actually divided, reaching from the top to the bottom, yet there are innumerable *imaginary*, which do as really produce effects by their preſſure, as if they were actually diſtinguiſhed. Theſe *imaginary* Pillars are repreſented in the firſt Schematiſm, one whereof is A E I N O P Q, the other B F K R T, and ſo forth.

THEOREM III.

There is a twofold Ballance, one Natural, another Artificial.

BY the *Artificial Ballance*, I underſtand that which the Mechanicks call *Libra*, which Merchants commonly uſe. By the *Natural Ballance* (which for diſtinctions cauſe
I ſo

I ſo nominat) I mean, *v. g.* a *Siphon* , or crooked Pipe,
wherein water naturally aſcends or deſcends, as high or low
in the one Leg, as in the other, ſtill keeping an evenneſs,
or likeneſs of weight.

THEOREM IV.

Fluid bodies counterpoiſe one another in the Ballance *of*
Nature, *according to their* Altitude *only.*

THis Theorem will appear afterwards moſt evident,
while we paſs through the ſeveral Experiments ; and
it is of ſpecial uſe for explicating ſundry difficulties that
commonly occur in the *Hydroſtaticks.* The meaning of it
is ſhortly this : while two Cylinders of Water are in the
oppoſite Scales of the *Natural Ballance,* they do not counterpoiſe one another according to their thickneſs : for
though the one Pillar of Water be ten times thicker, then
the other, and conſequently heavier, yet is it not able to
preſs up the other, that's more ſlender, and ſo lighter,
beyond its own hight : and therefore they weigh only according to their *Altitudes.*

THEOREM V.

In all Fluids there is a Preſſure.

Figure 1.

THis is true not only of the Elements of Air, and Water, while they are out of their own place (as they
ſpeak) but while they are in it. For Air and Water, being naturally indued with weight, the ſecond foot cannot
A 2

be

be under the firſt, unleſs it ſuſtain it: if this be, it muſt
neceſſarily be preſt with its burden. So this Water being
naturally a heavy body,the foot I cannot be under E, un‑
leſs it ſuſtain it, and be preſt with the burden of it ; the
foot N,being burdened with them both. From this Preſ‑
ſure, which is in Air, ariſeth a certain ſort of force, and
power , which may be called *Benſil*, by vertue whereof, a
little quantity of Air,can expand and ſpread out it ſelf,to a
very large quantity, and may by extrinſick force be redu‑
ced to that ſmall quantity again. Though this expanſive
faculty be evident in Air, yet it is ſcarcely diſcernable in
Water, unleſs it be in very deep parts, near the bottom,
where the Preſſure is great. This Preſſure is not of the
ſame Degree in all the parts, but is increaſed and augmen‑
ted, according to the deepneſs of the Air,and Water : for
the Air upon the tops of Mountains, and high places, is
thought to be of a leſs Preſſure, then in Valleys: and Wa‑
ter is of a leſs Preſſure, ten or twelve foot from the top,
then twenty or thirty. So is the Water N, under a far
leſs Preſſure, then the Water, P or Q.

THEOREM VI.
The preſſure of Fluids is on every ſide.
Figure 1.

THe meaning is, that Air and Water preſſeth not on‑
ly downward,but upward,not to the right hand only,
but to the left alſo,and every way.So the foot of water K,
not only preſſeth down the foot R, but preſſeth up the
foot F, yea preſſeth the foot I, and the foot L, with the
ſame weight. And the firſt imaginary ſurface , is as much
preſt

preſt up, by the water I K L M, as it is preſt down by the water E F G H. Upon this account it is, that when a Sphere, or Glob is ſuſpended in the middle of Water, or Air, all the points of their ſurfaces are uniformly preſt. After this manner, are our bodies preſt with the invironing Air, and the man that *dives*, with the ambient and invironing Water.

THEOREM VII·

All the parts of a Fluid in the ſame Horizontal Line, are equally preſt.

Figure 1.

THe meaning is, that the foot I, is no more preſt, then the foot K: neither is the foot L, more burdened, then the foot M. The reaſon is, becauſe each of theſe feet, ſuſtains the ſame weight: for E F G H are all of them, of the ſame burden: therefore all the parts of a Fluid in the ſame Horizontal ſurface, are preſt moſt equally. This holds true in Air, and Mercury, or in any other Liquid alſo.

THEOREM VIII·

The Preſſure of Fluids ſeem to be according to Arithmetical Progreſſion.

Figure 1.

THe meaning is, that if the firſt foot of Water, have one Degree of Preſſure in it, the ſecond muſt have only two, and the third muſt have only three, and ſo forth; which

which appears from the Schematiſm : for the firſt foot E,
having one Degree of weight, and the ſecond foot I, having of its ſelf as much, and ſuſtaining E, it muſt have two
Degrees, and no more. So the foot N, ſuſtaining two
Degrees of Preſſure from I and E, muſt have the weight
only of three Degrees, O of four, P of five. It's evident
alſo from Experience, for while by the Preſſure of Water,
Mercury is ſuſpended in a glaſs tub, we find, that as the
firſt fourteen inches of Water, ſuſtains one inch of Mercury, ſo the ſecond fourteen inches ſuſtains but two, and the
third, but three. But if the Preſſure were according to
Geometrical progreſsion, the third foot of Water ought to
ſuſtain four inches of Mercury, the fourth, eight; the fifth,
ſixteen, *&c.* which is contrary to Experience.

THEOREM IX.

In all Fluids there is a twofold weight, one Senſible,
the other Inſenſible.

THe firſt is common to all heavy bodies, which we
find in Water, while we lift a Veſſel full of it from
the ground. The *Inſenſible weight* of Water, and Air,
or of any other Fluid, can ſcarcely be diſcerned by the
ſenſes, though it be as real, as the former, becauſe the
Preſſure is uniform. By vertue of the ſecond, bodies naturally lighter than Water, are driven from the bottom to
the top, as *Cork.* So, a man falling into a deep Water,
goes preſently to the bottom, and inſtantly comes up again.
Here is a natural effect, which cannot want a natural cauſe;
and this can be nothing elſe, but the Preſſure of the Water, by vertue whereof he comes up, and yet he finds nothing

thing driving him up, or pulling him up. Therefore, there is in all Fluid bodies, an *Insensible* weight , as there is one *Sensible* ; seing the man that (perhaps) weighs seventeen Stone, is driven up fifteen or sixteen fathom by it. And it must be very considerable, and exceed the weight of the man, seing it is able to overcome such a weight. So are vapours and smoke driven upward by the *Insensible weight* of the Air, and by that same weight, do the Clouds swim above us.

THEOREM X.

The Insensible weight of Fluids, is only found by sense, when the Pressure is not uniform.

FOr understanding of this Proposition, I must suppose somethings that are possible , but not practicable. Put the case then, while a man opens his hand, the Air below were removed, he would scarce be able to sustain the weight of the Air, that rests upon the Palm above : or if the Air above were annihilated, he would not be able to bear down the weight that presseth upward. Or, while a *Diver* is in the bottom of the Sea, if it were possible to free any one part of his body from the Pressure of the Water, suppose his right arm, I doubt not, but the blood would spring out in abundance from his finger-ends : for the arm being free, and the other parts extreamly prest, the blood of necessity must be driven from the shoulder downward, with force, which cannot be without considerable pain. It is evident also, from the application of the *Cuppin-glass*, which being duely applied to a mans skin, causeth the Air to press unequally; the parts with-

out,

out, being more preſt, than the parts within, in which
caſe the unequal Preſſure cauſeth the pain, and ſo is found
by ſenſe.

THEOREM XI.

*A Cylinder of Water, or of any other Fluid body, loſeth of its
weight, according to its reclination from a* Perpendicular *poſition, towards an* Horizontal *or
levell ſcituation.*

FOr underſtanding of this, conſider that while a
Pipe full of Water ſtands perpendicular, the loweſt
foot ſuſtains the whole weight of the Water above it:
but no ſooner you begin to recline the Pipe from that Poſition, but aſſoon the Preſſure upon the loweſt foot grows
leſs; So that if the loweſt foot, in a perpendicular poſition, ſuſtained the burden of ten feet, it cannot ſuſtain
above five or ſix, when it is half reclined. A certain evidence whereof is this, the more a Cyilnder of Water is
reclined towards the Horizon, or Level, it takes the ſhorter
Cylinder of Water to counterpoiſe it, as is evident in *Siphons*. For, though the one Leg, be ſixteen inches long,
and the other but ſix; yet a Cylinder of Water ſix inches
long, will counterpoiſe a Cylinder of ſixteen. But this
cannot be, unleſs an alteration be made in the Preſſure.
For, how is it poſſible, that a Cylinder of Water can
ſometimes be in *æquilibrio* with a leſſer, and ſometimes
with a greater weight, unleſs the Weight, and Preſſure of
it, be ſometimes more, and ſometimes leſs? When I ſay
a Cylinder of Water loſeth of its weight by reclination,
it is to be underſtood only of the *Inſenſible Weight:* for
the

the *Sensible Weight* is unchangeable, seing it is alwayes a
Pillar of so many inches, or feet. Now the true reason,
why the Pressure upon the lowest foot grows less, is this;
the more the Pipe is reclined, the more weight of the
Cylinder rests upon the sides of the Pipe within; by
which means, the lowest foot is eased of the burthen, and
is altogether eased, when once the Pipe lyes Horizontal.

THEOREM XII.
*All motion in Fluids, is from the unequal Pressure
of the Horizontal surface.*

Figure 1.

FOr understanding this, I must distinguish a twofold motion in Fluids; one *common*, another *proper*, by vertue of the first, they incline, as all other heavy bodies, to
be at the center of the Earth. It is evident in the motion
of Rivers, which descend from the higher places to the valleys, even by vertue of that tendency they have to be at
the *center*. By vertue of the second, they incline to move
every way; not only downward, but upward, hither and
thither. This sort of motion is peculiar, and proper only
to Fluids; and it is that which is spoken of in this Theorem. I say then, that all motion in Fluids, is from the unequal Pressure of the Horizontal surface. For put the case
A, were more prest then B, *e.g.* with a stone, then surely as
the part A descends, the other part B will ascend, and so
will C and D rise higher too. Suppose next, the part A
were fred of the Pressure of the Air, then surely in the
same instant of time, would the part A ascend, and the
parts B C D descend. As this Proposition is true in order

B　　　　　　　　　　　　to

to the firſt and viſible ſurface A B C D, ſo it is true in order to the *imaginary* ſurtace I K L M ; for put the caſe the
ſpace I, were filled with a body naturally heavier then
Water, as lead or ſtone, then behoved that part of the ſurface to yeeld, it being more preſt, then the part of the ſame
ſurface K. Or if the ſpace K were filled with a body naturally lighter then water, as Cork, then ought the water R to aſcend, it being leſs preſt, then the water N or S.

THEOREM XIII.
*A body naturally heavier then Water, deſcends ; and
a body naturally lighter, aſcends.*
Figure 1.

FOr underſtanding of this, let us ſuppoſe the quadrat
ſpace E, to be filled with a piece of Lead or Iron. I
ſay then it muſt go down to I ; and the reaſon is, becauſe
the quadrat foot of Water I, is more preſſed then the
quadrat foot of Water K. To illuſtrat this, let us ſuppoſe that each quadrat foot of this Water weighs a pound,
and that the heavy body exiſting in E, weighs two
pound. If this be, the foot of Water I, muſt yeeld, ſeeing it is more preſt then K : upon the ſame account muſt
the Water N yeeld, and give way to the Stone, ſeeing
it is more preſt then R. For according to the twelfth
Theorem, *There cannot be unequal Preſſure upon a ſurface,
unleſs motion follow.*

For underſtanding the ſecond part, let us ſuppoſe the
ſpace R, to be filled with a piece of Cork, that is ſpecifically or naturally lighter then Water. I ſay then, it
muſt aſcend to the top B ; and the reaſon is, becauſe the
quadrat foot of Water K, is more preſt upward, then the
quadrat

quadrat foot of Water I, or L is : but this cannot be in
Fluid bodies, unless motion follow thereupon. I say, it
is more preſt up, becauſe R being lighter then N, or S,
it muſt preſs with greater force upon K, then S can do
upon L, or N upon I. It is ſtill to be remembred,
That Fluids preſſeth with as much ſtrength upward, as down-
ward, according to the ſixth Theorem; and that an Horizontal ſurface doth as really ſuffer unequal Preſſure from
below, as from above.

THEOREM XIV.

Bodies naturally lighter then Water, ſwim upon
the ſurface and top.

Figure 1.

THe reaſon of this Propoſition muſt be taken from
the nature of an *equipondium*, or equal weight. For
without doubt, there is a counter-ballance between the
Preſſure of the Water, and the weight of the body that
ſwims. To make this probable, let us ſuppoſe there were
a piece of Timber in form of a Cube, ſix inches thick every way, without weight. In this caſe, the under-ſurface
of that four-ſquar'd body, being applied to the ſurface of
the Water A, would ly cloſs upon it, as one plain Table
lyes upon the face of another, without any preſſure: and
it being void of weight, the part of the ſurface A, would
be no more burdened, then the next part B adjacent,
whence no motion would follow. Here is no *equipondium*,
or counter-ballance.

Secondly, let us ſuppoſe the ſaid body to acquire two
ounces of weight, then it follows, that it muſt ſubſide,
and ſink two inches below the ſurface A B C D; and that

ſo

ſo far, till it come by vertue of its new acquired weighr, to a counter-ballance with the Preſſure of the Water. Which Preſſure is nothing elſe, but as much force or weight, as is equivalent to the weight of Water, that is thruſt out of its own place, by the ſubſiding and ſinking of that body, two inches.

Thirdly, let us ſuppoſe the ſame body to acquire other two ounces of weight, then muſt it ſubſide other two inches. Laſtly, let us ſuppoſe that it acquires ſix ounces of weight, then it follows that the whole body ſinks, ſo far, I mean, till its upmoſt ſurface be in an *Horizontal line* with the ſurface of the Water A B C D. Here it ſwims alſo, becauſe the weight of it becomes juſt the weight of ſo much Water, as it hath put out of its own place. I ſay, it muſt ſwim, becauſe if the Water I, was able to ſuſtain the Water E, which is put from its own place, ſurely it muſt be able to ſuſtain that body alſo, that did thruſt it from its own place, ſeing both are of the ſame weight, namely ſix ounces. In this caſe, the body immerged, and the water wherein it is drowned, become of the ſame weight *ſpecifically*, ſeing bulk for bulk is of the ſame weight. To make this body *ſpecifically*, or naturally heavier then Water, and conſequently to ſink to the bottom, nothing is required, but to ſuppoſe that it acquires one ounce more of weight ; which done, it preſently goes down, I, being more burdened then K. Note by the way, a twofold weight in heavy bodies, one *individual*, the other *ſpecifick*, and that two bodies agreeing in *individual* weight, may differ in *ſpecifick* weight. So a pound of Lead, and a pound of Cork, agree *individually*, becauſe they are both 16. ounces: but they differ *ſpecifically*, becauſe the one is naturally heavier then the other.

THEOREM XV.

No Body that flots above Water, even though its upper surface be level with the surface of the Water, can ever be made to swim between the top and the bottom.

Figure 1.

FOr clearing this Proposition, let us suppose F to be a four-square piece of Timber, of the same *specifick* and natural weight with Water, and consequently its upper surface to be level with the surface of the Water A B C D. I say then, if it be prest down to R, it shall arise thence, and never rest till it be where it was, namely in F. The reason seems to be this, because the four-squar'd body of Water R, is really heavier, then the four-squar'd piece of Timber F. If this be true, it follows of necessity, that it must ascend: for if the Timber existing in R, be lighter then the Water R, the Water T must be less prest, then the Water O, or the Water V; whence (according to the twelfth Theorem) *motion must follow.* Again, if the Timber R, existing in the Water R, be lighter then the same Water is, then must the Water K, be more prest up then the Water I, or L; whence yet, according to the same Theorem, *motion must follow.* If it be said, that the Timber F, is of the same weight with the Water R, because, it being equal in weight with the Water F, which it hath thrust out of its own place, it must also be equal in weight to the Water R, seeing F and R being of the same dimensions, are of the same weight. There is no way to answer this difficulty, unless I say the four-squar'd body of water R, is really and truly heavier then the four-squar'd body of Water F. The
reason

reafon feems to be, becaufe the Water R, is under a greater Preffure, then the Water F ; and by vertue of this
greater Preffure, there are really *moe parts* of Water in it,
then in F ; therefore it muft be heavier. Even as there are
far moe parts of Air,in one cubick foot near the *Earth*,then
in fix or feven near the *Atmofphere*. Hence it is, that
a pint of Water taken from the bottom of the Sea, fourty
fathom deep, will be heavier, I mean in a ballance, then a
pint taken from the furface. Take notice, that when the
veffel is once full at the bottom, the orifice muft be clofely
ftopped, till it come to the top : otherwife the parts that
are compreffed at the bottom, namely by the weight of
the fuperiour parts, relaxes themfelves, before they come
to the top.

THEOREM XVI.
*It is not impofsible for a body to be fufpended between
the* furface *and the* bottom.

Figure 1.

FOr underftanding this, fuppofe F to be a four-fquare
piece of Timber, which though it will not reft but at
the furface, A B C D, yet may be made to go down of its
own accord, and reft at T, namely, by making it fo
much heavier, as the Water T is heavier then the Water
F. To know this difference, which is not very practicable ;
the Cube of Water T, muft be brought from its own place,
under the fame degree of Preffure it hath, and put into the
Scale of a Ballance, and weighed with the Cube of Water
F, put into the other Scale. Now if the Water T, be
half an ounce heavier, then the Water F, then to make
the Timber F hing in T, it muft be made half an ounce
heavier. There feems to be reafon for it alfo ; for if a
Cube

Cube of Timber resting in the space T, be just the weight of the Water T, the *imaginary* surface O T V, is no more prest, then if T were Water, and so it cannot go downward : neither can it go upward, seing the under part of the Water R, is no more prest up by the Timber T, then if the space T were filled with Water. If it be said, according to this reasoning, a Stone may be suspended in a deep Water, between the top and the bottom, which is absurd. I answer, such a thing may happen in a very deep Water : For put the case a Cube of Lead twelve inches every way, were to go down twelve thousand fathom , it is probable, it would be suspended before it came to the ground. For coming to an *imaginary* surface far down, where the Pressure is great, a Cube of Water twelve inches thick there, may be as heavy (even *specifically*) as the Cube of Lead is, though the Lead be ten times heavier *specifically*, then any foot of VVater at the top. If Water suffer compression of parts, by the superiour burden ; it is more then probable, that the second foot of Water burdened with the first, hath moe parts in it, then are in the first, and the third moe, then in the second, and so forth ; and consequently, that the second is heavier, then the first, and the third heavier, then the second. Now, if this be, why may not that foot of Water, that hath sixty thousand foot above it, by vertue of this burden, be so comprest, that in it may be as many parts, as may counter-ballance a Cube of Lead twelve inches every way ? If then, that *imaginary* surface, that is sixty thousand foot deep, be able to sustain the said foot of VVater, which perhaps weighs twenty pound, why may it not likewise sustain the Lead, that is both of the same dimensions with it, and weight ?

Hence

Hence it is, that the Clouds do ſwim in the Air, by vertue of a counter-ballance: And we ſee, which confirms this
Doctrine, that the thinneſt and lighteſt are alwayes far-
theſt up; and the thickeſt and blackeſt, are alwayes far-
theſt down.

THEOREM XVII.

*The lower the parts of a Fluid are, they are the heavier,
though all of them be of equal quantity and dimenſions.*

Figure 1.

THis follows from the former, which may appear a
Paradox, yet it ſeems to be true: for though the
Water Q at the bottom, be of the ſame dimenſions with
the Water E at the top, yet it is really heavier, which
happens (as I ſaid) from the ſuperiour Preſſure. It is clear
alſo from this, namely the Cube of Timber E, which
ſwims upon the ſurface, being thruſt down to Q, comes
up to the top again, which could not be, unleſs the Water Q, were heavier then the Water E. I ſuppoſe the Water E, and the Timber E, to be exactly of the ſame *ſpecifick* weight, and conſequently the ſurface of the Timber,
to ly Horizontal with B C D. Now the reaſon, why the
Timber aſcends from Q to E, is no other then this, namely that the one Water is heavier then the other; for the
under part of the Water P, being more preſt up with the
Timber exiſting in Q, then with the Water Q it ſelf, it
muſt yeeld and give way to the aſcent: for if the Cube of
Timber exiſting in Q, were as heavy as the Water Q it ſelf,
it would no more preſs upon P, or endeavour to be up, then
the Water Q does.

THEO-

THEOREM XVIII.
A heavy body weighs leß in Water, then in Air.
Figure 1.

THis is easily proven from experience; for after you have weighed a stone in the Air, and finds it two pound, and an half, take it, and suspend it by a threed knit to the scale of a ballance, and let it down into the Water, and you shall find it half a pound lighter. The question then is, why doth it lose half a pound of its weight? I answer, the stone becomes half a pound lighter, because the surface of Water on which it rests, sustains half a pound of it: For put the case a stone were resting in R, that weighed two pound and an half in the Air, it behoved to weigh but two pound in this Water; because the Water T sustains half a pound of it. For if this Water T be able to sustain the Water R, that weighs half a pound, it must be also able to sustain half a pound of the stone, seing half a pound of stone is no heavier, then half a pound of Water. Note, that when a heavy body is weighed in Water, it becomes so much lighter exactly, as is the weight of the Water it thrusts out of its own place.

THEOREM XIX.
A heavy body weighs less nigh the bottom of the Water, then nigh the top thereof.
Figure 1.

FOr clearing this proposition, I must suppose from the 17. Theorem, that the lower the parts of Water be, be, they are the heavier, though all of them be of equal dimensions. If then the lowest foot Q be heavier, that is, have moe parts in it, then the foot N, it of necessity follows, that a stone suspended in Q, must be lighter then while it is suspended in N or I. Because, if a stone be lighter in Water then in Air, as is said, even by as much, as is the weight of the bulk of Water, that the bulk of the stone expells, then surely it must be lighter in the one, then in the other place; because suspended in Q, it expells moe parts of Water, then while it is suspended in N or I. For example, let us suppose the Water N, to weigh eight ounces, and the Water Q to weigh nine; then must the stone suspended in Q, weigh less by an ounce, then suspended in N, seeing as much is deduced from the weight of the stone, as is the weight of the Water it expells: but so it is, that it thrusts nine ounces of Water out of its own place in Q, and but eight in N or I; therefore it must be one ounce lighter in the one place, then in the other. This may be tried, with a nice, and accurat ballance, which will bring us to the knowledge of this, namely how much the foot of Water Q is heavier, then the Water N or O.

THEOREM XX·

One part of a Fluid, cannot be under compression,
unless all the parts next adjacent, be under
the same degree of Pressure.

Figure 1.

THis proposition may be proven by many instances: for when the Air of a *Wind-gun*, is reduced to less quantity by the Rammer, all the parts are most exactly of the same *Bensil*. So is it in a Bladder full of wind. It's true,

true, not only in order to this artificial Preſſure, but in order to the natural Preſſure, and *Benſil* of the Air likewiſe.
For the Air within a parlour, hath all its parts, under the
ſame degree of natural compreſſion : ſo is it with the parts
of the Air, that are without, and immediatly under the
weight of the *Atmoſphere*. Its evident alſo in the parts of
Water : for the foot of Water R, cannot be under Preſſure, unleſs the Water S, and N, be under the ſame degree of it. Though this be true of Fluids, while all the parts
lye in the ſame Horizontal ſurface, yet to ſpeak ſtrictly, it
will not hold true of the parts ſcituated under divers ſurfaces ; for without queſtion, the foot of VVater T, muſt
be under four degrees of Preſſure, if the VVater R, be
under three. And if the Air in the loweſt ſtory of a building, be under ſix degrees of *Benſil*, the Air in the higheſt
ſtory muſt be under five. If a man would diſtinguiſh *Me-
taphyſically*, and ſubtilly, he will find a difference of this
kind, not only between the firſt, and ſecond fathom of Air,
neareſt to the Earth, but between the firſt, and ſecond foot;
yea, between the firſt and ſecond inch, and leſs, much more
in Water, as to ſenſe. However it be, yet the Theorem holds true ; for we find no difference ſenſible, between the compreſſion of Air in this room ; and the compreſſion of Air in the next room above it, no not with the
Baroſcope, or *Torricellian Experiment*, that diſcerns ſuch
differences accurately. I judge it likewiſe to be true, in
order to the next adjacent parts of Fluids of different kinds;
for while a ſurface of Mercury, is burdened with a Pillar of
Water, or a ſurface of Water, with a Pillar of Air, whatever degree of weight and Preſſure, is in the loweſt parts
of theſe Pillars, the ſame is communicated entirely, to the
ſurfaces, that ſuſtains them. So then, there is as much

force

force and power, in the ſurface of any Water, as there is
Weight and Preſſure, in the loweſt foot of any Pillar of
Air, that reſts upon it: otherwiſe, the ſurface of Water
would never be able to ſupport the ſaid Pillar : for a ſurface of ſix degrees of force, can never be able to ſuſtain a
a Pillar of Air, of eight, or ten degrees of weight.

THEOREM XXI.

*The Preſſure of Fluids, may be as much in the leaſt
part, as in the whole.*

Figure 1.

THis Theorem may ſeem hard, yet it can be made manifeſt, by many inſtances : for albeit the quantity
of Air, that fills a Parlour, be little in reſpect of the whole
Element, yet ſurely, there is as much Preſſure in it, as in
the whole ; becauſe Experience ſhews, that the *Mercurial
Cylinder* in the *Baroſcope*, will be as well ſuſtained in a
Chamber, as without, and under the whole *Atmoſphere*
directly ; which could not be, unleſs the ſmall portion of
Air, that's in this Parlour, had as much Preſſure in it, as
in the whole Element. Beſides this, it will be found in a
far leſs quantity : for though the *Baroſcope* were incloſed,
and impriſoned ſo cloſs, within a ſmall Veſſel, that the
Air within, could have no communion with the Air without, yet the Preſſure of that very ſmall quantity, will
ſuſtain 29 inches of Mercury, and this will come to paſs,
even though the whole Element of Air were annihilated.
This Propoſition is likewiſe evident in order to the Preſſure of the Water : for put the caſe, the *Baroſcope*, whoſe
Mercurial Cylinder is 29 inches, by the Preſſure of the

Air ;

Air; were sent down to the bottom of a Sea 34. foot deep, within a Veffel, as a Hogs-head, and there exactly inclofed, that the VVater within, could have no commerce with the VVater without, yet as well, after this fhutting up, as before, other 29. inches would be fuftained, by the Preffure of this imprifoned VVater, which proves evidently, that there is as much Preffure in one Hogs-head full of VVater, at the bottom of the Sea, as in the whole Element of VVater, above, or about: for an Element of VVater never fo fpacious, if it exceed not 34. foot in deepnefs, can fuftain no more Mercury, then 29. inches by its Preffure. Yea, though the Veffel with the *Baro-fcope*, and imprifoned VVater in it, were brought above to the free Air, yet will the VVater retain the fame Preffure, and will *de facto* fuftain 29. inches of Mercury, provided the Veffel be kept clofs. It is therefore evident, that as much Preffure may be in one fmall quantity of VVater, as in the whole Element, or Ocean. 'Tis to be obferved, that this Theorem is to be underftood chiefly of the lower parts of Fluids; feing there cannot be fo much Preffure in the VVater P, as in the VVater Q; for in effect, there is as much Preffure in the VVater Q, as is in the whole VVater above it, or about it. From this Theorem, we fee evidently, that the Preffure, and *Benfil* of a Fluid, is not to be meafured, according to its bulk, and quantity, feing there is as much *Benfil* in one foot, nay, in one inch of Air, as is in the whole Element, and as ftrong a Preffure in one foot of VVater, or lefs, as there is in the whole Ocean : therefore the greateft quantity of Air, hath not alwayes the greateft *Benfil*, neither the greateft quantity of VVater, the greateft Preffure. But this will appear more evident afterwards.

THEO-

THEOREM XXII.

The Preſſure, and Benſil *of a Fluid, is a thing, really diſtinct from the* natural weight *of a Fluid.*

Figure 1.

THis may be eaſily conceived ; for as in ſolid bodies, the *Benſil*, and *natural weight*, are two diſtinct things, ſo is it in Air, and Water, or in any other Fluid. The ~~weight~~ of a *Bow*, is one thing, and the *natural weight* of it, is another. The weight of the Spring of a *Watch*, and the *Benſil* of it, are two diſtinct things. The weight (perhaps) will not exceed two ounces : but the *Benſil* (may be) will be equivalent to two pound. Though theſe may illuſtrate, yet they do not convince: therefore I ſhall adduce a reaſon, and it's this. The *natural weight* of a Fluid is leſs, or more, as the quantity is leſs or more ; but it is not ſo with the Preſſure, becauſe there may be as much Preſſure in a ſmall quantity, as in a great, as is evident from the laſt Theorem, therefore they may be different. The firſt part of the Argument is manifeſt, becauſe there is more weight in a gallon of Water, then in a pint. A ſecond reaſon is, becauſe a Fluid may loſe of its preſſure, without loſing of its weight. This is evident from the Schematiſm, for if you take away the four foot of Water E F G H, and conſequently make the four Pillars ſhorter, the foot of Water Q becomes of leſs Preſſure, but not of leſs Weight, ſeeing the quantity ſtill remains the ſame : at leaſt, the loſs of weight is not comparable, to the loſs of Preſſure. I ſay, it becomes of leſs Preſſure, becauſe there is a leſs burden above it. Thirdly, the Preſ-

ſure

sure and *Benfil* may be intended, and made stronger, without any alteration in the weight: so is the *Benfil* of Air, within a Bladder, made stronger by heat, without any alteration, in the weight of it. Likewise, the Pressure of the foot of Water Q, may be made stronger, by making these four pillars higher, without any alteration, at least considerable, in the weight; for it still remains a foot of water, whatever be the hight of the pillars above it. Lastly, the weight of a Fluid is *essential* to it, but the Pressure is only *accidental*; because it is only generated, and begotten in the inferiour parts, by the weight of the superiour, which weight may be taken away.

THEOREM XXIII.

Though the Benfil *of a Fluid, be not the same thing formally* with the weight, yet are they the same *effectively.*

THis proposition is true in order to many other things, besides Fluids: for we see that the *Sun*, and *Fire*, are *formally* different, yet they may be the same *effectively*; because the same effects, that are done by the heat of the *Sun*, may be done by the heat of the *Fire*. So the same effects, that are produced by the *weight* of a Fluid, may be done by the Pressure, and *Benfill* of it. Thus, the Mercurial Cylinder in the *Torricellian Experiment*, may be either sustained by the *Benfil* of the Air, or the *weight* of it. By the *Benfil*, as when no more Air, is admitted to rest upon the stagnant Mercury, then three or four inches, the rest being secluded, by stopping the orifice of the Vessel. By the *weight* of it, as when an intire Pillar of Air, from the top of the Atmosphere, rests upon the face of the stagnant
Quick-

Quickfilver. It is also evident in a *Clock*, which may be
made to move, either by a weight of Lead, or by the force,
and power of a Steel *Spring*.

THEOREM XXIV.

The surfaces of Waters, are able to suftain any weight what-
foever, provided that weight prefs equally, and uniformly.

Figure 1.

THis is evident, becaufe the imaginary furface of
VVater O T V X, doth really fupport the whole
fixteen Cubes of VVater above it, yea, though they were
fixteen thoufand, And the reafon is, becaufe they prefs
moft equally, and uniformly. VVhat I affirm of the imaginary furface, the fame I affirm, of the firft and vifible.
For let a plain body of lead, never fo heavy, be laid upon
the top of the VVater A B C D, yet will it fupport it,
and keep it from finking, provided it prefs uniformly all
the parts of that furface. It is clear alfo, from the fubfequent Theorem.

THEOREM XXV.

The furfaces of all Waters whatfoever, fupport as much
weight from the Air, as if they had the weight of thirty
four foot of Water above them, or twenty nine
inches of Quick-filver prefsing them.

THis Propofition is evident from this, that the Preffure of the Air, is able to raife above the furface of
any Water, a Pillar of Water thirty four foot high. For,

put

put the case there were a *Pump* fourty foot high , erected among stagnant Water , and a *Sucker* in it , for extracting the internal Air, a man will find, that the Water will climb up in it four and thirty foot ; which *Phænomenon* could never happen , unless the surface of the stagnant Water, among which the end of the *Pump* is drowned , were as much prest with the Air , as if it had a burden of Water upon it thirty four foot high. The second part is also evident, because if a man drown the end of a long Pipe, in a Vessel with stagnant Quick-silver , and remove the Air that's within the Pipe by a *Sucker* , or more easily by the help of the *Air-pump* , he will find the Liquor to rise twenty nine inches, above the surface below, which thing could never come to pass, unless the Pressure of the Air, upon the surfaces of all Bodies , were equivalent to the Pressure and weight of twenty nine inches of Quick-silver.

THEOREM XXVI.

All Fluid Bodies have a sphere of Activity , to which they are able to press up themselves , or another Fluid, and no further, which is less or more, according to the altitude of that pressing Fluid.

Figure 2.

FOr understanding this Proposition , let us imagine G H C D to be a Vessel , in whose bottom, there are five inches of Mercury E F C D. Next, that above the stagnant Mercury, there are thirty four foot of Water resting, namely A B E F. Lastly , that upon the surface of the said Water, there is resting the Element of Air G H A B, whose top G H , I reckon to be about

D   six

ſix thouſand fathom above A B. Beſides theſe, let us
imagine, that there are here three Pipes, open at both ends,
the firſt whereof C A G, having it's lower orifice C,
drowned among the ſtagnant Mercury E F C D, goeth
ſo high, that then pper orifice goeth above the top of the
Air G H. The ſecond, whoſe lower orifice I, is only drowned among the Water A B E F, reaches to the top of the
Air likewiſe. The third, whoſe open end K, is above
the ſurface of the VVater A N B, and hanging in
the open Air, goeth likewiſe above the Atmoſphere.
Theſe things being ſuppoſed, we ſee that no Fluid can,
by its own proper weight, preſs any part of it ſelf, higher
then it's own ſurface, ſeing the ſtagnant Mercury E F C D,
cannot preſs it ſelf within the Pipe C G, higher then E.
Neither can the VVater A B E F, preſs it ſelf higher
within the Pipe I L, then the point N. Laſtly, neither
can the Air G H A B, preſs it ſelf within the Pipe K M,
higher then M. But when one Fluid preſſeth upon another, as the VVater A B E F, upon the Mercury E F C D,
then doth the ſaid Mercury aſcend higher than it's own
ſurface, namely from E to O, which point is the higheſt,
to which the thirty four foot of VVater A B E F, can
raiſe the Mercury, which altitude, is twenty nine inches
above the ſurface E I F. But if a ſecond Fluid be ſuperadded, as the whole Air G H A B, then muſt the Mercury, according to that new Preſſure, riſe by proportion;
ſo riſes the Mercury from O to P, other twenty nine
inches. By this ſame additional weight of Air, the Water riſes thirty four foot in the Pipe I L, namely from N
to R. Now, I ſay, the outmoſt and higheſt point, to
which the Element of Air G H A B can raiſe the Mercury, is from O to P; for by the Preſſure of the Water

ter A B E F, it rises from E to O. And the highest
point, to which the said Air can raise the VVater, is from
N to R. The reasons of these determinate altitudes,
must be sought for, from the altitudes of the incumbing
and pressing Fluids: for as these are less or more, so is the
altitude of the Mercury, and of the VVater within the
Pipes more or less. The hight therefore of the Mercury E O, is twenty nine inches, because the deepness of
the pressing water A B E F is thirty four foot. And the
hight of the VVater N R, is thirty four foot, because the
hight of the Air G H, above A B, is six thousand
fathom, or thereabout. And for the same reason, is the
Mercury O P twenty nine inches.

THEOREM XXVII.
A lighter Fluid, is able to press with as great
burden, as a heavier.
Figure 2.

THis Proposition is true, not only of VVater in respect of Mercury, but of Air in respect of them
both: for albeit Air be a thousand times lighter then
VVater, yet may it have as great a Pressure with it, as
VVater; as is evident from this second Schematism,
where by the Pressure of the outward Air G H A B,
twenty nine inches of Mercury O P are supported, as
well as the twenty nine inches E O, by the Pressure of
the VVater A B E F. So doth the same Air, sustain the
thirty four foot of VVater N R, which are really as heavy, as the twenty nine inches of Mercury O P. Now,
if the weight of the Atmosphere, be equivalent to the
weight

weight of thirty four foot of Water, or of twenty nine
inches of Mercury, 'tis no wonder to see Water press
with as great weight as Mercury; which is likewise clear
from this same Figure, where by the Pressure of the Water A B E F, twenty nine inches of Mercury E O are
suspended, as truly as the Mercury C E, within the lower
end of the Pipe, is supported by the outward invironing
Mercury. The reasons of these *Phenomena*, are taken
from the altitudes of the pressing Fluids: for though a
Body were never so light, yet multiplication of parts makes
multiplication of weight; which multiplication of parts
in Fluids, must be according to altitude: for multiplication of parts according to thickness and breadth will not
do it. Observe here, that if as much Air, as fills the Tube
between N and L, were put into the scale of a Ballance,
it would exactly counterpoise the thirty four foot of Water
N R, poured into the other scale. Item, that as much
Water as will fill the Tube between E and A, is just
the weight of the Mercury E O. Lastly, that as much
Air as will fill the Pipe, between O and G, is just the
weight of the Mercury O P.

THEOREM XXVIII.
The Pressure of Fluids, doth not diminish, while you
subtract from their thickness, but only, when
you subtract from their altitude.
Figure 1.

FOr understanding this, let us look upon the first Schematism, where there are four Pillars of Water. Now
I say, though you cut off the three Columes of Water,
upon the right side, yet there shall remain as much Pressure,

sure, in the quadrat foot of VVater Q, as was, while these were intire. But if you cut off from the top, the VVater E F G H, then presently an alteration follows, not only in the lowest parts, nigh to the bottom, but through all the intermediat parts: for not only the VVater Q loseth a degree of its Pressure, but the VVaters P and O suffer the same loss. This Theorem holds true likewise in order to the Element of Air. For if by *Divine Providence*, the Air should become less in Altitude, than it is ; then surely, the *Bensil* of the ambient Air, that we breath in and out, should be by proportion weakned also. And contrariwise, if the Altitude became more, then stronger should the *Bensil* be here, with us, in the lowest parts: both which would be hurtful to creatures, that live by breathing. For if the Altitude of the Air, were far more then it is, our bodies would be under a far greater Pressure, which surely would be very hurtful. And upon the other hand, if the Altitude of the Air, were far less then it is. we should be at a greater loss; for then, by reason of the weak *Bensil*, we would breath indeed, but with great difficulty.

THEOREM XXIX.

A thicker Pillar of a Fluid, is not able to press up a slenderer, unless there be an unequal Pressure.

Figure 3.

For understanding this, let us suppose this third Schematism to represent a vessel with VVater in it, as high as A B, among which is thrust down to the bottom, the Pipe G H, open at both ends. I say then, the two thicker

Pillars

they are heavier, becaule they are thicker. I anlwer, they
are truly heavier, for the Pillar of Air F B apart, will be
thrice as heavy, as the flender Pillar of Air I G. But, if
you reckon the Pillar of Air E A, upon the left hand,
both together, will be fix times heavier, then the Air
I G: yet are they not able, either feverally, or con'unct-
ly, to prefs up the Water H I, higher then I, or the
Air I G, higher then G. For folving this difficulty, I
muft fay conform to the fourth Theorem, that Fluid Bodies, counterpoifeth one another, not according to their
thicknefs, and *breadth*, but according to their *altitude* only: therefore, feing the flender Pillar of Air I G, is as
high, as either F B, or E A, it cannot be preft up by
them. For by vertue of this equal hight, all the three
prefs equally and uniformly, upon the furface of Water
A B; and therefore according to the twelfth Theorem,
there can be no motion. But if fo be, the Pillar F B,
were higher then the Pillar I G, then furely would
the Water H I, be preft up; for in fuch a cafe, there is
an unequal Preffure. Or if the Pillar I G, were higher
then the Pillar F B, then furely would the Water I H be
preft down, there being again an unequal Preffure : the
Water within the Pipe, being more burdened then the
Water about the Pipe. In a word, there's no more difficulty here, then if the Pipe were taken away : in which
cafe, there would be but one Pillar of Air, refting upon
the furface of Water A B. If it be faid, the Pipe being
thruft down, makes of one Pillar, three diftinct ones, and
conse-

consequently a formal counter-ballance, or mutual suspension. Be it so, yet because all three press uniformly, there can be no motion.

THEOREM XXX.

Fluids press not only according to perpendicular Lines, but also according to crooked Lines.

Figure 4.

For proving this Proposition, let us suppose A B C D, to be a large Vessel full of VVater, as high as A N B, and a little Vessel lying within it, near to the bottom, close above at M, but with an open orifice downward, as G, and having other two passages going into it, upon the right, and left side, as E O, and F P. Now, I say, the Pressure of the VVater, is not only from N to M, in a straight line downwards, but from E to O, and from F to P, by crooked lines. Nay, put the case this Vessel had no passage in to it, but by a *Labyrinth*, or entry full of intricate windings, yet the Pressure will be communicated, thorow all these, even to the middle of it; and which is more, the VVater H or I, within the Vessel, would be under the same degree of Pressure, with the VVater E or L, without; or with the VVater K or P. And which is stranger, let us suppose both the entries E and F stopped, and nothing remaining open, but the hole G, which I judge no wider, then may admit the hair of ones head, yet thorow this small hole, shall the Pressure be communicated, to the parts of the Water within. In as high a degree, as if the upper part of the Vessel E M L, were cut off, to let the Pressure come down directly. What is our

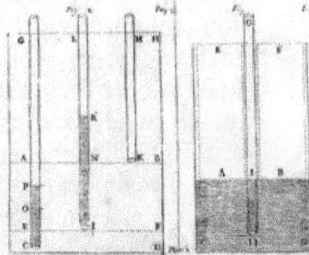

A	B	C	D
E	F	G	H
I	K	L	M
N	R	S	Y
O	T	V	X
P			
Q			

in order to Water, the same is true in order to Air, or Mercury, or any other Fluid. For, though a house were built never so clofs, without door, or window, yet if there remain but one fmal hole in it, the Preffure of the whole Atmofphere, fhall be tranfmitted thorow that entrie, and fhall reduce the Air within the houfe, to as high a degree of Benfil, as the Air without.

THEOREM XXXI.

The Preffure , and Benfil of a Fluid , that's in the Low-eft foot, is equivalent to the weight of the whole Pillar above.

Figure 5.

For underftanding this Propofition, let us fuppofe E F to be the loweft foot of a Pillar of Air, cut off from the reft, and inclofed in the Veffel E F, fix inches in Diameter, or widenefs, and twelve inches high. Now I fay, the *Benfil* and Preffure, that's in that one foot of Air, is exactly of as great force and power, as is the weight of the whole Pillar of Air, from which it was cut off. Let A B be that Pillar of Air, which I fuppofe is fix inches thick , and fix thoufand fathom high. Take it, and weigh it in a Ballance, and fay it weighs 500 pound, yet the Preffure, and *Benfil*, that's in the Air E F, is of as much force : and if the one be of ftrength by its weight, to move, *v. g.* a great *Clock*, the other by its *Benfil*, will be of as much. This propofition is true alfo in order to Water. For put the cafe E F, were the loweft of 34 foot of Water: in it will be found as much Preffure, and force, as will be equivalent to the weight of the whole thirty three foot , from which

it

it was cut off. But here occurreth a difficulty; for if the Preſſure, and *Benſil* of the foot of Air E F, be equivalent to the weight of the whole Pillar of Air A B, which weighs 500 pound, then muſt the ſlender Pillar of Air C D, that's but two inches in diameter, be as heavy weighed in a ballance, as the thicker Pillar A B, which is abſurd. I prove the connexion of the two parts of the Argument thus: as the Benſil of the Air G H, is to the Benſil of the Air E F, ſo is the weight of the Pillar C D, to the weight of the Pillar A B: but ſo it is, that the Benſil of the Air G H, is equal in degree to the Benſil of the Air E F, according to the Theorem 21. Where it's ſaid, that the Preſſure of Fluids may be as much, in the leaſt part, as in the whole: therefore the Pillar C D, and the Pillar A B, muſt be of equal weight, when both are weighed together in the oppoſite ſcales of a Ballance, which is falſe, ſeing the one is far thicker, and ſo heavier then the other. There's no way to anſwer this objection, but by granting the Air G H, and E F, to be equal in *Benſil,* and yet the two Pillars unequal in weight, becauſe according to the 22 Theorem, the *Benſil* of a Fluid is one thing, and the *natural weight* is another.

THEOREM XXXII.

In all Fluids there is a Pondus *and a* Potentia, *a* weight *and a* power, *counterpoiſing one another, as in the* Staticks.

THat part of the *Mathematicks*, which is called *Staticks*, is nothing elſe, but the *Art of weighing heavy Bodies*; in which, two things are commonly diſtin
E guiſhed

guiſhed, *viz.* the *pondus* and the *potentia*, the *weight* and
the *power*. 'Tis evident, while two things are counterpoiſing one another in the oppoſite ſcales of a Ballance,
as *Lead* and *Gold*, the one being the *pondus*, the other the
potentia. The ſame two are as truly found in the *Hydroſtaticks* : for while the Mercurial Cylinder is ſuſpended
in the *Torricellian Experiment*, by the weight of the Air,
the one is really the *pondus*, the other the *potentia*. Or
while into a *Siphon*, with the two orifices upward, Water
is poured, there ariſes a counterpoiſe, the Water of the
one Leg counter-ballancing the Water of the other; this
taking the name of a *pondus*, the other the name of a *potentia*. 'Tis evident alſo, while a ſurface of Water, ſuſtains a Pillar of Water, this being the *pondus*, that the *potentia* : Or, while a ſurface of Water ſuſtains a Pillar
of Air, the Pillar of Air being the *pondus*, and the ſurface
of Water the *potentia*. Or, while a ſurface of Quick-
ſilver ſuſtains a Pillar of Water or Air; the ſurface is the
power, and either of the two is the *pondus*, or *weight*, as
you pleaſe.

THEOREM XXXIII·

Fluid Bodies can never ceaſe from motion, *ſo long
as* the pondus *exceeds the* potentia, *or the*
potentia *the* pondus.

THIS is a ſure Principle in the *Hydroſtaticks*, which
will appear moſt evident; while we paſs thorow the
ſubſequent Experiments, I ſhall only now make it appear
by one inſtance, though afterwards by a hundred. In the
Torricellian Experiment, lately mentioned, 'tis obſerved,
that

that though the Pipe were never so long, that's filled with Mercury, yet the Liquor subsides, and falls down alwayes till it come twenty nine inches above the surface of the stagnant Mercury below. The reason whereof is truly this, so long as the Mercury is higher then the said point, as long doth the *pondus* of it exceed the *potentia* of the Air; therefore the motion of it downward can never cease, till at last by falling down, and becoming shorter, it becomes lighter, in which instant of time, the motion ends, both of them being now in *equipondio*, or in evenness of weight.

THEOREM XXXIV.

When two Fluids of different kinds are in æquilibrio together, the height of the one Cylinder is in proportion to the height of the other, as the natural weight of the one is to the natural weight of the other.

FOr understanding this Theorem, we must consider, that when two Cylinders of the same kind, as one of Water with Water, or as one of Mercury with Mercury, are counterpoising one another, both are of the same altitude, because both are of the same natural weight. But when the two are of different kinds, as a Cylinder of Air with Mercury, or as a Cylinder of Air with Water, or as a Cylinder of Water with Mercury, then it will be found, that by what proportion, the one Liquor is naturally heavier or lighter, then the other, by that same proportion, is the one Cylinder higher or lower then the other. For example, because Air is reckoned 14000 times lighter then Quick-silver, therefore the Pillar of Air that counter-

poiseth

poiſeth the Pillar of Quick-ſilver in the *Torricellian Ex-*
periment, is 14000 times higher. The one is 29 inches,
and therefore the other is 406000 inches : which will
amount to 33833 foot, or about 6766 fathom, counting
five foot to a fathom. And becauſe Air is counted 1000
times lighter then Water, therefore the Pillar of Air that
ſuſtains the Pillar of Water is 1000 times higher. The
hight of Water by the Preſſure of the Air is 34 foot,
and therefore the hight of the Air is a thouſand times 34
foot. And becauſe Water is reckoned 14 times lighter
than Mercury, therefore you will find, even by experience,
that the Pillar of Water, that counterpoiſes the Pillar of
Mercury, is 14 times higher. For if the Mercury be
ten inches, the Water will be exactly 140. If it be 29
inches, the Water will be thirty four foot. The reaſon
is evident, becauſe if one inch of Mercury be as heavy
naturally as 14 inches of Water, it follows of neceſſity,
that for making of a counterpoiſe, to every inch of Mercury, there muſt be 14 of Water, and theſe in altitude,
each one above another.

Hydro-

Hydrostatical

EXPERIMENTS,

For demonstrating the wonderful
Weight, *Force*, and *Pressure* of the
Water in its own Element.

EXPERIMENT I.

Figure 6.

IN explicating the *Phenomena* of the *Hydrostaticks*, and in collecting speculative, or practical conclusions from them, I purpose to make choise of the plainest, and most easie Experiments, especially in the entry, that this knowledge, that's not very common, and yet very useful, may be communicated to the meanest capacities. For, if at the first, any mystical, or abstruse Experiments, should be proposed with intricate descriptions, they would soon discourage, and at last hinder the ingenuous Reader from making progress

greſs. For, if a man do not take up diſtinctly, the Experiment it ſelf firſt, he ſhall never be able to comprehend next the *Phenomena*, nor at laſt ſee the inferences of the concluſions. Next, though ſome of the *trials* may ſeem obvious, yet they afford excellent *Phenomena*, by which many profound ſecrets of Nature are diſcovered. And if that be, 'tis no matter what kind they be of. Then, the grand deſign here, is not to multiply bare, and naked Experiments; for that's a work to no purpoſe, for it's like a foundation without a ſuperſtructure: but the intention is, not only to deſcribe ſuch and ſuch *things*, but to build ſuch and ſuch Theorems upon them, and to infer ſuch and ſuch concluſions, as ſhall make a ſtately building, and give a man in a ſhort time a full view of this excellent Doctrine.

For the firſt Experiment then, prepare a Veſſel of any quantity, as A B C D, near half full of Water, whoſe ſurface is M H. Prepare alſo two Glaſs-pipes, the one wider, the other narrower, open at both ends, which muſt be thruſt down below the Water, firſt ſtopping the two upper orifices E and F. This done, open the ſaid orifices, and you ſhall ſee the Water aſcend in the wider to G, and in the narrower to H. Now, the queſtion is, What's the reaſon, why the Water did not aſcend, the orifices E and F, being ſtopped, and why it aſcends, they being opened? To the firſt part I anſwer, the Water cannot aſcend, becauſe the imaginary ſurface of Water L K is equally and uniformly preſt: for with what weight the outward Water M L, and H K preſs the ſaid ſurface, with the ſame weight, doth the Air within the two Pipes preſs it. To the ſecond part I anſwer, the Water aſcends, becauſe the ſame ſurface (the orifices E and F being opened) is unequally preſt:

preſt : for the outward Water M L, and H K, preſs it
more, then the Air within the Pipes do. The difficulty
only is, why it is equally preſt, the orifices E and F being ſtopped, and why it is unequally preſt, the ſaid orifices
being once opened. To unlooſe the knot, I muſt ſhew the
reaſon, why the Air within the Pipes, preſs the ſurface
L K, with as great a burden, as the outward Water preſs
it. For underſtanding this, you muſt know, that when
the orifice I is thruſt down below the Water, there ariſeth a ſort of debate between the lower parts of the Water, and the Air within the Pipes, the Water ſtriving to
be in at I, and the Air ſtriving to keep it out : but becauſe
the Water is the ſtronger party, it enters the orifice I,
and cauſeth the Air retire a little up, one fourth part, or
ſixth part of an inch, above I, and no more, which is a
real compreſſion it ſuffers. For the orifice E being ſtopped, hinders any more compreſſion, than what is ſaid ; in
which inſtant of time the debate ends, the Air no more
yeelding, and the Water no more urging ; by which means
the Air having obtained a degree of *Benſil*, more then ordinary, by the Preſſure of that little quantity of Water,
that comes in at I, preſſeth the part of the imaginary ſurface, it reſts upon, with as great weight, as the outward
Water preſſeth the parts it reſts upon. But when the
orifice E is opened, the outward water M L, and H K,
preſs the imaginary ſurface L K more, than the Air within the Pipe can do. And the reaſon is, becauſe by opening the orifice above, the internal Air, that ſuffered a
degree of *Benſil* more then ordinary, preſently is freed,
and conſequently becomes of leſs force, and weight ; which
the Water finding, that hath a little entered the orifice I,
inſtantly aſcends to G, it being leſs preſſed, then the Wa-

ter

ter without the Pipe. Now the reaſon, why it aſcends
no higher then G, is taken from the equal Preſſure of the
Body that reſts upon the ſurface M G H : For, aſſoon as
it comes that length, all the parts of the horizontal Plain
of Water, is uniformly preſt with the incumbing Air,
both within the Pipe, and without the Pipe. The Water
in going up, cannot halt mid-way between I and G, for
then there ſhould be an unequal Preſſure in Fluids without motion, which is impoſſible; for the Water is ſtill
ſtronger then the Air, till once it climb up to G.

From this Experiment we ſee firſt, that in Water there
is a Preſſure and Force; becauſe having opened the orifice E, which is only *cauſa per accidens* of this motion,
the Water is preſt up from I to G. We ſee ſecondly,
that Fluid Bodies, can never ceaſe from motion, till there be
an equal Preſſure among the parts, which is evident from
the aſcent of the Water from I to G, which cannot halt
in any part between I and G, becauſe of an unequal Preſſure, till it once climb up to G. We ſee thirdly, that
Fluid Bodies do not ſuſtain, or counterpoiſe one another
according to their *thickneſs* and *breadth*, but only according to their *altitude*; becauſe there is not here any proportion between the ſlender Pillar of Water H K within
the Pipe, and the outward Water that ſuſtains it, I
mean as to the *thickneſs*; therefore 'tis no matter, whither
the Glaſs Tubs be wider or narrower, that are uſed in
counterpoiſing Fluid bodies one with another. And this
is the true reaſon, why 'tis no matter, whither the Tub of
the *Baroſcope* be a wide one, or a narrow one, ſeing the Air
doth not counterpoiſe the Mercury, according to *thickneſs*, that's to ſay, neither the thickneſs of the ambient
Air that ſuſtains, nor the thickneſs of the Mercury that

is

is suftained, are to be confidered ; but only their *altitudes*.
'Tis true, the element of Air is fourteen thoufand times
higher , then the Mercurial Cylinder, yet there is a
certain and true proportion kept between their heights ;
fo that if the element of Air , fhould by *divine providence*
become higher or lower, the height of the Mercury would
alter accordingly.

<hr>

EXPERIMENT· II·
Figure 6.

TAke out of the Water, the wide Pipe E G I, and
ftopping the orifice I , pour in Water above at E,
till the Tub be compleatly full.   Having done this, thruft
down the ftopped orifice I to the bottom of the Veffel,
and there open it , then fhall you fee the Water fall down
from E to G, and there halt.   The reafon is taken from
unequal Preffure ; for the Tub being full of Water from
E to I, that part of the *imaginary* furface, upon which
the Pillar of Water refts, is more burdened than any other
part of it , namely more then L or K ; therefore feing
one part is more burdened than another , the Cylinder of
Water that caufeth the burden, muft fo far fall down, till
all the parts be alike preft, in which inftant of time , the
motion ceafeth.   This leads us to a clear difcovery of the
reafon , why in the *Barofcope* , the Mercury falls from the
top of the Tub of any height, alwayes to the twentieth and
ninth inch , above the ftagnant Quick-filver.   For example, fill the Pipe N Q, which is fixty inches high
with Mercury , and opening the orifice Q, the Liquor
fhall fall out, and fall down from N, till it reft at R, which

F   is

is twenty nine inch above the open orifice Q. The reafon is the fame, namely unequal Preffure, feing one part of the *imaginary* furface of Air X S, upon which the Cylinder of Mercury ftands, is more burthened then the other next adjacent: therefore, fo long and fo far muft the Mercury fubfide and fall down, till the part Q, upon which the Bafis of the Pillar refts, be no more burthened, than the reft of the parts; in which inftant of time, the motion ceafeth, and there happeneth an equal ballance, between the Silver within the Tub, and the Air without. If it be faid, I fee a clear reafon, why the outward Water M L, ought to fuftain the inward G I, but cannot fee, why the outward Air T Z S and V R X, ought to fuftain the inward Mercury R X: neither do I fee a reafon, why it fhould halt at R, as the Water refts at G. I anfwer, though fenfe cannot perceive the one, as evidently as the other, yet the one is as fure as the other. For taking up the reafon why it halts at R, 29 inches above X, you muft remember, from the 25 Theorem, that the Preffure of the Air upon Bodies, is equivalent to the weight of 34 foot of VVater perpendicularly, or 29 inches of *Quick-filver*. The Pillars of Air then T Z S, and V R X, being as heavy each one of them, as two Pillars of Mercury, each one of them 29 inches high, it follows of neceffity, that the Mercury within the Tub, muft be as high as R. 'Tis no wonder to fee the *Silver* halt at R, provided R X, and Z S, were two bulks of Mercury, environing the Pipe, as the outward VVater environs the wider and narrower Pipe. Neither ought any to wonder, when the *Silver* falls down, and refts at R, nothing environing the Pipe but Air, feing the Preffure of the Air is equivalent to the weight of 29 inches of *Quick-filver*.

This

This Experiment is eaſily made : take therefore a ſlender Glaſs-pipe of any length, beyond 30 inches, open at both ends ; but the lower end Q, muſt be drawn ſo ſmall by a flame of a Lamp, that the entry may be no wider, than may admit the point of a ſmall needle, or the hair of ones head. Then ſtopping the ſaid orifice, pour in Mercury above at the orifice N, till the Pipe be compleatly full. Next, cloſe the ſaid orifice with wet Paper, and the pulp of your finger; and opening the lower orifice, you ſhall find, (which is very delightful to behold) the Mercury ſpring out, like unto a ſmall ſilver threed, and falling down from the top N, ſhall reſt at R, the motion ceaſing at the narrow orifice Q. This ſhews evidently, that there is not need alwayes of ſtagnant Mercury, for trying the *Torricellian Experiment*; but only when the mouth of the Pipe below is wide : for being narrow, the *ſilver* runs ſlowly out, and conſequently ſubſides ſlowly above, and coming down ſlowly to R, there reſts. But when the mouth is wide below, the *ſilver* falls down ſo quickly, that it goes beyond R, before it can recover it ſelf, which recovery would never be, unleſs there were ſtagnant Mercury to run up again.

From what is ſaid, we ſee firſt, that when one part of a ſurface of Water or Air, is more burthened than another, the burthened part preſently yeelds, till it be no more burthened than the other. This is clear from the falling down of the Water from E to G, which cannot be ſupported by the part I, becauſe more burthened than the reſt. We ſee ſecondly, that the element of Air, reſts upon the ſurfaces of all bodies with a conſiderable weight ; otherwiſe it could not ſuſtain the Water, before it fall down from E to G : for if it did not reſt upon the ſurface

MH,

M H, with weight, the Water could never be ſuſpended; ſeing the application of the finger to the orifice E, is only the *accidental cauſe* of this ſuſtentation. We ſee thirdly, that according to the difference of *natural weight*, between two Fluids, ſo is the proportion of *altitudes* between two of their Cylinders: thereſore Air being reckoned 14000 times lighter then Mercury, it followes that the Cylinder of Mercury ſuſtained by the Air, muſt be 14000 times lower and ſhorter, than the Cylinder of Air that ſuſtaines it; which appears from this experiment to be true, ſeeing by the Preſſure of the Air, which is thought to be about 7000 fathom high, 29 inches of Mercury is ſupported between R and X. In a word, if Air be naturally 14000 times lighter than Mercury, which is very probable; then muſt the altitude of it, commonly called the *Atmoſphere*, be fourteen thouſand times, nine and twenty inches, that is 406000, or of feet 33833.

EXPERIMENT III.
Figure 6.

WHile the outward, and inward Water are of the ſame altitude, withdraw the inward Air E G by ſuction, or by any other device you think fit, and you will find the Water riſe as high as E, which I ſuppoſe to be 34 foot above M G H. The ſame *Phenomenon* happens, in taking the Air out of the narrow Pipe F K. The reaſon is ſtill unequal Preſſure; for in removing the Air, that's within the Pipe, the part of the ſurface M, and the part H, remaines burthened, while the part G is freed of its burden: therefore this part of the ſurface, being liberated

rated of its burden, that came down through the Pipe, instantly rises, and climbs up as far, as the outward Air resting upon M and H, can raise it, which is to E 34 foot: for the Preſſure of the Air upon the ſurfaces of all Waters, according to the 25 *Theorem*, being equivalent to the weight of 34 foot of Water, muſt raiſe the ſaid Water in the Pipe 34 foot. You do not wonder, why it riſes from I to G, as in the firſt experiment; no more ought you to wonder, why it riſes from G to E, ſeing the weight of the Air, doth the ſame thing, that 34 foot of Water reſting upon the ſurface M H, would do.

From this experiment we ſee firſt, that the Preſſure of the Air, is the proper cauſe of the motion of Water, up thorow *Pumps* and *Siphons*, or any other inſtrument, that's uſed in Water-works of that kind; for if the weight of the Air, reſting upon the ſurface M H be the cauſe, why the Water climbs up from G to E, the ſame muſt be the cauſe, why the ſtagnant Water followes the *Sucker* of the *Pump*, while it's pulled up. And the ſame is the cauſe, why Water aſcends the Leg of a *Siphon*, and is the cauſe, why motion continues after ſuction is ended. We ſee ſecondly, that every Preſſing Fluid hath a *Sphere of activity*, to which it is able to raiſe the Fluid, that is preſſed. This is evident in this experiment, becauſe the Preſſure of the Air reſting upon M H, is able to raiſe the Water, the hight of E in the wide Pipe, and the hight of F in the narrow, and no further, even though the ſaid Pipes were far longer: and this altitude and higheſt point is preciſely 34 foot between Air and Water. We ſee thirdly, that 'tis all one matter, whether *Pumps* and *Siphons* be wider or narrower, whether the tub of the *Baroſcope* be, wherein the Mercury is ſuſpended, of a large Diameter, or of a leſſer

Diameter,

Diameter. This is alſo evident from the ſame experiment; ſeing there is no more difficulty in cauſing the Water aſcend the wide Pipe, than in cauſing it aſcend the narrow one. And the reaſon is, becauſe the preſſing Fluid reſpects not the preſſed Fluid, according to its *thickneſs* and *breadth*; but only according to its *altitude*. Therefore its as eaſie for the Air, to preſs up Water through a *Pump* four foot in Diameter, as to preſs it up through a *Pump*, but one foot in Diameter.

EXPERIMENT IV.
Figure 7.

THis Schematiſm repreſents a large Veſſel full of Water, whoſe firſt and viſible ſurface is D E H K. The ſecond, that's imaginary is, L I, ſix foot below it. The third of the ſame kind, is M G, ſix foot lower. The fourth, is N F O, ſix foot yet lower. The laſt, and loweſt, is A B C. There are here alſo four Tubs, or rather one Tub under four divers poſitions, with both ends open. After this Tub D A is thruſt below the Water, till it aſcend, as high as D in it, lift it up between your fingers, till it have the poſition of the ſecond Pipe E F, and then you ſhall ſee, as the orifice of the Pipe aſcends, the Cylinder of Water fall out by little and little, until it be no longer than E F. Again, lift it further up, till it have the poſition of the Pipe H G, then ſhall you find the Cylinder of Water become yet ſhorter. Laſtly, if it be ſcituated, as the Pipe K I, the internal Water becomes no longer than K I. The reaſons of theſe *Phenomena* are the ſame; namely unequal Preſſure; for the Orifice A being lifted up as high

high as F, it comes to the imaginary surface N O, which
is not under so much Pressure, as the other is; therefore
one part of it being more burdened, than another, namely
the part upon which the Cylinder of Water rests, it presently yeelds, and suffers the Cylinder to become shorter,
and lighter, till it become no heavier, then is proportionable to its own strength. To make this reason more evident, it is to be noted, that no surface of Water is able to
support a Cylinder higher then its own deepness, that is to   *Maxime*
say, if a surface be 40 foot deep, it is able to sustain a Cylinder 40 foot high, and no more: therefore the surface
N O, being but 18 foot deep, it cannot sustain a Cylinder 24 foot long: for if that were, then the *Potentia*,
should be inferiour to the *Pondus*, which is impossible in
the *Hydrostaticks*. In effect, it were no less absurdity,
then to say, 18 ounces are able to counterballance 24. For
a second trial, lift up the same Pipe higher, till it acquire
the position of the Tub G H; in this case, the Cylinder
of Water within it, becomes yet shorter, even no longer,
than G H. The reason is the same, namely unequal Pressure; for when a Cylinder of Water 18 foot high, comes
to rest upon this surface, that is but 12 foot deep, it makes
one part of it more burdened then another; therefore the
part that is more prest, presently yeelds, and suffers the
Cylinder to fall down, till the *Pondus* of it, become equal
to its own *Potentia*. For the last trial, lift up the Tub,
till it acquire the position of the Pipe K I: in this case, the
Water within it becomes no longer then K I, the surface
L I, that is but six foot deep, not being able to sustain a
Cylinder 12 foot high.

From this Experiment we see first, that in all Fluid Bodies there is a Pressure, which is more or less, according to
the

the deepness of that Fluid; this is evident from the four
several surfaces; there being more Pressure and force in the
lowest A B C, then in the next N O; and more in this,
then in the surface M G; and more in this, then in L I.
We see secondly, that in all Fluids, there is a *Pondus* and
a *Potentia*; which two are alwayes of equal force, and
strength; the *Potentia* is clear and evident in the surface,
by supporting the Pillar; which Pillar is nothing else, but
the *Pondus* supported. And that they are alwayes of equal
strength, is most evident also; for when you endeavour
to make the *Pondus* unequal to the *Potentia*, in making a
surface 18 foot deep, to support a Pillar 24 foot high, they
of their own accord become equal; the Pillar becoming
shorter, and suitable to the strength of the surface that
sustains it. We see thirdly, that 'tis impossible for one
part of the same Horizontal surface, to be more burdened
then another: for when you endeavour to do it, by setting
a longer Pillar upon it, the part burdened instantly yeelds,
till it be no more prest, then the next part to it. We see
fourthly, that the inequality, that is between the *Pondus*
and the *Potentia* in Fluids, is the proper cause of the motion of Fluids. For when you endeavour to make a surface 30 foot deep, sustain a Pillar 40 foot high, this inequality is the true cause, why the Pillar subsides, and falls
down, and why the surface yeelds, and gives way to it.
And this inequality is the true cause, why the motion of
Water thorow *Siphons* continues. For understanding this,
you must conceive a *Siphon*, to be nothing else, but a crooked Pipe with two legs, the one drowned among Water,
the other hanging in the open Air. The use of it is, for
conveying Wine or Water from one Vessel to another,
which is easily done by suction. Now after suction is ended,

ed, the motion of the Water continues, till the furface
become lower, then the orifice out of which it runs. The
true reafon then, why the Water flows out, is the inequality between the *Potentia* of the Air, and the *Pondus* of the
VVater; the *Pondus* being ftronger then the *Potentia*.
For in Air as in VVater, we muft conceive Horizontal
furfaces; and thefe furfaces to be endowed with Preffure
and force, as are the furfaces of VVater. Now when the
leg of a *Siphon* is hanging in the Air, it muft reft upon one
furface or another, and confequently the VVater in it,
muft reft upon the fame furface. If the *Potentia* of the
furface be ftronger, then the *Pondus* of the VVater; the
VVater is driven backward, which alwayes comes to pafs,
when the orifice is higher, then the furface of the VVater
of the Veffel, among which the other leg is drowned. If
the *Potentia* of the furface of that Air, be of equal power
and ftrength, with the *Pondus* of the VVater, the VVater
goeth neither backward, nor forward, but ftands in *equilibrio*: this happens, when the orifice is neither higher, nor
lower, than the furface of the VVater in the Veffel. But
if the *Potentia* of the furface of the Air be weaker, than the
Pondus of the VVater; in this cafe, the Air yeelds, and
fuffers the VVater to run out, even as a furface 30 foot
deep, yeelds to a Pillar of VVater 40 foot high. The fame
inequality is the reafon, why VVater climbs up the Pump;
why VVater climbs up a Pipe, when a man fucks with his
mouth. Before fuction, the *Potentia* that's in the furface
of VVater, among which the end of the Pipe is drowned,
is of equal force with the *Pondus* of the Pillar of Air, that
comes down thorow the Pipe, or Pump; but affoon as a
man begins to fuck, the faid Pillar of Air becomes lighter;
and the VVater finding this, prefently afcends. The fame

G   is is the reason, why the Mercury falls down to 29 inches in the *Baroscope*, and no further: for as long as the *Pondus* of the Pillar of Mercury, exceeds the *Potentia* of the surface of Air, so long doth the motion continue; and when both are become equal in force, the motion ceaseth. VVhen the Glass-tub is 40 inches long, and filled with Mercury, and inverted after the common manner, you are endeavouring as it were, to cause a surface 29 inches deep, sustain a Pillar 40 inches high, which is utterly impossible in Fluids. It is judged by many a wonder to see the deflux of the Mercury in the *Baroscope*; but in effect, there's no more cause of admiration in it, than to see the Cylinder of Water grow shorter, by lifting the Pipe up from one surface to another.

From this Experiment, we see the true reason, why the Mercurial Cylinder of the *Baroscope* becomes shorter and shorter, according as a man climbs up a mountain with it. For at the root of the hill, the surface of Air, that sustains the Pillar of Mercury, is of greater force, than the surface at the middle part: and this is stronger than any surface at the top. The Pipe therefore being carried up from one surface to another, the Mercury in it, must subside, and fall down, even as the Water falls down, and becomes shorter, by lifting the Pipe from the surface A B C D to the surface N O. And as the whole VVater would fall down, if the orifice I, were lifted above the surface D E H K, so if the *Baroscope* could be carried so high, till it came above the top of the Air, the whole Mercurial Cylinder would surely fall down. And as by thrusting down the said Pipe to the bottom of the Vessel again, as the Pipe D A, the VVater ascends in it; so by bringing down the *Baroscope* to the earth again, the whole 29 inches would rise again.   EXPE-

EXPERIMENT V.
Figure 8.

Ill the Veſſel **A D G H** with VVater to the brim, Next, thruſt down the open orifice of the Tub **D A**, to the bottom, and you ſhall ſee the VVater aſcend in it, as high as **D**, according to the fiſt experiment. When this is done, recline the ſaid Pipe, till it ly as **B E**, and you ſhall find the Pipe, compleatly full of VVater. Next, erect the ſame Tub again as **D A**, and you ſhall ſee the Cylinder of VVater fall down, and become ſhorter, as at fiſt. For ſalving this *Phenomenon*, and ſuch like, I muſt ſuppoſe this VVater to be 50 inches deep, and the Tub **I A**, and **B E** 90 inches long: and the ſaid Tub in reclining, to deſcribe the quadrant of a Circle **F E G**. Now the queſtion is, why there being but 50 inches of Water in the Tub, while erected, there ſhould be 90 in it, when it is reclined? Secondly, why there ſhould be 90 inches of Water in the Tub **B E**, and but 50 in it, when it ſtands Perpendicular, as **D A**? If you reply, becauſe there are 90 inches *in recta linea* between the point **B**, and the point **E**, and but 50 between **A** and **D**. But this will not anſwer the caſe; becauſe, if you ſtop the orifice **E**, with the pulp of your Finger, before it be erected, you will find the Tub remain full of VVater, even while it ſtands Perpendicular; and fall down, when the orifice is opened. Or, while the Tub ſtands Perpendicular, ſtop the orifice **I**, and recline it as **B E**: yet no more Water will be found in it, than 50 inches: but by unſtopping the ſaid orifice, the VVater climbs up from **R** to **E**, and becomes 90 inches. Now, what's the reaſon,

why

why it runs up from R to E , and why it falls down from
I to D? I answer then, the VVater must run up from R
to E, because of the inequality, that's between the *Pondus*
of the Cylinder B R , and the *Potentia* of the surface of
VVater A B C , that supports the said Cylinder. For
underſtanding this, know, while the Tub is erected , there
is a perfect equality , between the weight of the Pillar A
D, and the force or *Power* of the surface that sustains it, seing a surface 50 inches deep, supports a Pillar 50 inches
high. But aſſoon as the Tub is reclined, there ariſes ane
inequality between the saids two parties , the *Pondus* of
the Cylinder becoming now leſs than before. If you say
the quantity of the VVater is the same, namely 50 inches,
in the reclined Tub , as well as in the Perpendicular. I
grant the quantity is the same , but the weight is become
leſs. Now the reaſon, why the same individual VVater, is
not ſo heavy as before , is this ; there are 40 ounces of
it , supported by the sides of the Tub within ; which
were not, while the Tub was erected: for in this poſition, the whole weight of the Cylinder reſts upon the surface:
but while the Tub is reclined, the said surface is eaſed , and
freed of 40 ounces of it ; this 40 , reſting and leaning upon
the sides of the Pipe within. The surface then , finding
the said Cylinder lighter now than before, inſtantly drives
it up from R to E, 40 inches. And likewiſe, when the
reclined Pipe is made Perpendicular, the Water falls down
from I to D, becauſe of the inequality, that's between
the *Pondus* of the Pillar , and the *Potentia* of the surface ;
this surface 50 inches deep, not being able to support a Pillar 90 inches high, for if this were, then one part , ſhould
be more burthened than another, which is impoſſible.

    It is to be obſerved , that by how much the more, the
Tub

Tub is reclined from a Perpendicular, towards the horizontal surface **A B C** , by so much the more growes the inequality, between the *Pondus* and the *Potentia*, and that according to a certaine proportion. Hence is it, that the Tub being reclined from 60 degrees to 50 , there arises a greater inequality between the *Pondus* of the Cylinder, and the *Potentia* of the surface, than while it is reclined from 70 to 60 : and more yet in moving from 50 to 40 , than in moving from 60 to 50, and so downward, till it be horizontal, in which position, the whole *Pondus* is lost. And contrariwise, while the Pipe is elevated, the *Pondus* begins to grow ; and growes more , being lifted up from 10 to 20, than from 1 to 10 : and yet more in travelling from 20 to 30, than from 10 to 20, and so upwards, till it be Perpendicular, in which position, the Cylinder regaines the whole *Pondus* and weight , it had. This proportion is easily known , for its nothing else , but the proportion of *Versed Sines* upon the line F B ; for according to what measure, these unequal divisions become wider , and wider from 90 to 1 , according to the same proportion does the *Pondus* of the Cylinder become less and less: and contrariwise, according to what proportion the said divisions become more and more narrow from 1 to 90 , according to the same measure and rate , does the *Pondus* of the Cylinder become greater and greater.

From this experiment we see first , that two Cylinders of Fluid bodies, differing much in quantity , may be of the same weight: because though the Cylinder B E 90 inches long, be far more in quantity , than the Cylinder D A, that's but 50, yet both of them are of the same weight, in respect of the surface that sustaines them. If it be said, the one is really heavier , than the other , notwithstanding of all.

all this. I anſwer, it is ſo indeed, in reſpect of the *Libra*, or
Artificial Ballance, that we commonly uſe in weighing of
things: but it is not ſo in reſpect of this *Natural Ballance*, if
I may ſo ſpeak, wherein Fluid bodies are onely weighed after this manner. We ſee ſecondly a clear ground for ſetting down the ninth Theorem, namely, that in all Fluid
bodies a twofold weight may be diſtinguiſhed, one *Senſible*,
another *Inſenſible*: becauſe the *Senſible* weight of the Cylinder of Water B E, remaines ſtill the ſame, even though
it ſhould be reclined to G; for take it out, and weigh it
in a Ballance, it will be as heavy the one way as the other.
But it is not ſo with the *Inſenſible* weight; ſeeing the Tub
begins no ſooner to recline, but aſſoon it begins to diminiſh, and grows leſs. This *Inſenſible* weight is nothing
elſe, but the *ſenſible* weight conſidered after another manner.
For look upon the weight of the Pillar of Water B E, as
it weighs in a pair of Scales, it is then *Senſible*, and weighs ſo
many ounces, and cannot be more or leſs: but look upon
it in reference to the *Potentia* of the ſurface, that ſuſtains it,
it is then *Inſenſible* as to us: for though a man ſhould put
his hand below the Water, and endeavour to find the
weight of the ſaid Pillar, yet he ſhall not find it, though
that part of the ſurface upon which it reſts, doth really
(if I may ſo ſpeak) find the weight of it. And as it is
Inſenſible, ſo is it ſometimes more, and ſometimes leſs,
according as the Tub is elevated, or reclined: now theſe
two being put together, gives a very probable ground for
this diſtinction. We ſee thirdly, that the *Pondus* or weight
of Fluids, doth not only preſs according to Perpendicular
lines, but according to lines falling obliquely upon the
imaginary ſurface; ſo doth the weight of the Pillar of Water B E, preſs the ſurface A B C. We ſee fourthly, that
Fluid

Fluid Bodies, do nevertheless not touch; according to *Monsieur* only : for put the cold, the Pipe B F, were ten times wider than it, yet will the further fall in the Water in it, provided the Pipe keep still the same position of *Altitude*, namely 30 degrees : the reason seems to be this, for it the *parts* of the *Pillar* become more in Diameter, it necessarily requires a larger part of the surface to rest upon, which larger part is really stronger than the lesser part, so will be forced downwards. From this Experiment we see lastly, in consequence, why the Mercurial Cylinder in the *aereoparentes tip*, and still the empty space, when the Pipe is inclined, and why it runs down, when the Tub is inclined again. In effect, the reason is that there is an inequality between the *Ponder* of the Quick-Silver, and the *Parents* of the Surface of the Air, so when the Tub begins to rise, the *Ponder* begins to rest upon the side of the Tub which, by which means the *Parents* of the Surface finding the surface less, instantly clouds up the Stagnant Mercury to supply that loss, being two Fluids cannot counterpoise one another, which they be upon the one. And consequently, observe the Tub begins to be erected, the *Ponder* of the Mercury begins to grow, and so overcomes the *Parents* of the Surface, till by falling down it can do no more.

EXPERIMENT VI.

Figur 9.

This Schematim represents a Vessel full of Water, whose flat and visible surface is H I K, the bottom, which is imaginary, is E F G; the third, A B C D, Besides

Besides these three in Water, conceive a fourth in the Air, above the Water, namely L M N. Upon this aërial surface, rests the orifice M, of the Tub T M, open above. Upon the surface E F G, is standing the mouth F, of the Pipe S F. And upon the surface A B C D, stands the Pipe R B, open at both ends. After the orifice B is drowned below the VVater, you will find the Liquor rise from B to H. Then close with the pulp of your Finger the mouth R, and lift the Pipe so far up, till it have the Position of the Pipe S F ; and you shall see the VVater hing in it between F and O. Lastly, bring the said orifice compleatly above the VVater, till it have the position of the Tub T M; yet shall the VVater still hing in it, as M P. The first question is, what sustains the VVater I O ; for the part F I, is sustained by the ambient VVater? I answer, it cannot be the pulp of the Finger closing the orifice S ; for though, by taking away the Finger, the VVater O I falls down, and by putting to the Finger, it is keeped up, yet this proves not the pulp of the Finger to be the principal, and immediat cause. I say then, the VVater O I is suspended by the weight of the incumbing Air, resting upon the surface H I K. For understanding this, consider, as I said before, 25. Theorem, that the Pressure of the Air upon all Bodies, is just equivalent to the weight of 34 foot of VVater. Hence then is it, that if the Air be able to sustain a Pillar of VVater, 34 foot high, it must be able to sustain the short Pillar O I, that exceeds not four foot. The second question is, whether the part F, be equally burthened with the part E, or G ; for it would seem not, seing the VVater O I F, is but four foot high ; whilest upon E or G is resting, not only more then a foot of VVater to the top H I K, but
the

the whole weight of the *Atmoſphere* upon the ſaid top
is reſting, which is equivalent to the burden of 34 foot of
VVater. I anſwer, there's more to be conſidered, than
that four foot of VVater, which in it ſelf is but of ſmall
burden, therefore to this we muſt add the weight of the
Air between O and S, within the Pipe (remember that
the orifice S is ſtopped with the pulp of the Finger) which
in effect will be as heavy as 31 foot of VVater. Put the
caſe then, F, to be one foot below the firſt ſurface H I K,
and the VVater O I to be three foot, then ought the
Air O S, to have the weight of 31 foot, becauſe the ſurface E F G is able to ſupport a Pillar of 35 foot. This
I prove, becauſe the part E, *de facto*, ſuſtains 35 foot, becauſe the Air above is equivalent to 34 foot of it, and
there is a foot of VVater between it and the top, namely
between E and H. The third queſtion is, how it comes
to paſs, that the Water ſtill remains in the Pipe, after the
orifice M is brought above the ſurface of the Water; for
there is here no ſtagnant Water guarding it, as guards the
orifice F. I anſwer, that the *baſe* M, of this Pillar of
Water P M, as really reſts upon the horizontal ſurface of
this Air L M N, as a Cylinder of Braſs or Timber reſts
upon a plain Marble Table, and after the ſame manner.
Remember that the orifice T is ſtopped all this time, with
the pulp of the Finger. If it be ſaid, that the part M, is
more burdened then the part N, ſeing it ſuſtains four foot
of Water, which the part N ſupports not, and the Air
P T within the Pipe alſo, which is of as much *Benſil* and
Preſſure, as the Air N Y is of. For clearing of this difficulty, conſider, that the Pillar P M is ſhorter now than
before; for the orifice M coming up from D, ſome inches
of Water falls out, as will be found by experience. Sup-

H

pose then, that of four foot, six inches fall out ; if this
be, then the inclosed Air between P and T, must be six
inches longer, if this be, then of necessity the Bensil of it
must be proportionably remitted and slackened: whence
follows by *Metaphysical* necessity, that it cannot burden the
Water P M, with as much weight as it had, and consequently the surface of Air cannot be so much burdened.
It must then be no more bu dened with them both together, than it is with the single Pillar of Air Y N. If then
the Water P M, be three foot and an half, the weight
of the enclosed Air T P, must be exactly the weight of
thirty foot of Water and an half.

From this experiment, we see first the Pressure of the
Air, for by it the Water O I is suspended, and by the
same pressure is the Water P M suspended. We see secondly, that in Air, there is a power of dilating it self, and
that this dilatation never happens, without a relaxation of
the *Bensil*. We see thirdly, that one Fluid cannot sustain
another, unless the *Potentia* of the one, be equal to the
Pondus of the other, as is clear from the Aërial surface, that
cannot sustain the whole four foot of Water, but suffers
six inches of it to fall out, that the *Pondus* of the rest, and
the Air above it, may become equal to its own *Potentia*.
We see fourthly, that Fluid Bodies have not only a power
of pressing downward, but of pressing upward likewise:
as is clear from the Water O I, that's suspended by the
Air pressing down the surface of Water H I K. It presseth upward also, while it supports the Water P M. This
Experiment also answers a case, namely, whether or not, it
is alwayes needful to guard the orifice of the Tub of the
Baroscope with stagnant Quick-silver? I say then, it is not
alwayes needful, provided the orifice be of a narrow
                                            diameter ;

diameter; for experience tells, that while it is such, the Mercury will subside, and halt at 29 inches above the orifice, though no stagnant Mercury be to guard. In making this trial, the orifice must be no wider, than may admit the point of a needle. Or suppose it to have the wideness of a *Tobacco-pipe*, yet will the Mercury be suspended, though the end be not drowned among stagnant *Quick-silver*, even as the Water P M, is kept up without stagnant Water about it. For trial of this, you must first let the end of the Pipe, be put down among stagnant Mercury, and after the Cylinder is fallen down to its own proper altitude, lift up the Pipe slowly, till the orifice come above the surface, and you will find, provided you do not shake the Pipe, the Cylinder to be suspended after the same manner, immediatly by the Air, as the Water P M is.

EXPERIMENT VII.
Figure 10, 11.

TAke a Vessel of any quantity, such as A B C D E, and fill it with VVater. And a Glass-pipe, such as G F D, of 15 or 20 inches long, of any wideness, close above, and open below. Before you drown the open end among the VVater, hold the Glass before the fire, till it be pretty hot, and having put it down, you will see the VVater begin to creep up till it come to F, where it halts. The question now is, what's the reason, why the VVater creeps up after this manner, 10 or 12 inches above the surface A B? I answer, the heat having rarified the Air within, and by this means, having expelled much of it, and the Air now contracting it self again with cold, the

VVater afcends, being preft up with the weight of the incumbing Air, refting upon the furface of Water A B.
There is here furely an inequality between a *Pondus* and a
Potentia, that muft be the caufe of this motion. I judge then
the inequality to confift between the weight of the Air
within the Pipe, and the furface of Water CDE. To explicate this, I muft fuppofe the Pipe to be thruft down cold;
in this cafe, little or no Water can enter the orifice D. And
the reafon is, becaufe the *Pondus* of the Air within the
Glafs, is equal to the *Potentia* of the furface C D E. But
when the Pipe is thruft down hot, much of the Air having
been expelled by the heat, and now beginning to be contracted by cold, the *Pondus* of the Air becomes unequal to
the *Potentia* of the furface, and therefore this, being the
ftronger party, drives up the Air within the Glafs, till by
this afcent, the *Pondus* of the Air G F, and the *Pondus*
of the Water F D together, become equal to the *Potentia*
of the furface C D E, that fuftains them. For a fecond
trial; bring a hot coal near to the fide of the Glafs, between G and F, and you will find the Water to creep down
from F toward the furface A B; and if it continue any
fpace, it will drive down the whole Water, and thruft it
out at D. To explicate this, I muft fuppofe that heat,
by rarifying the Air within the Glafs, intends and increaf-
eth the *Benfil* of it, and the *Benfil* being now made ftronger, there muft arife an inequality between the *Pondus* of
the faid Air, and the *Potentia* of the furface C D E; the
Air then, being the ftronger party, caufeth the furface to
yeeld.

By comparing this Experiment with the former, we fee
a great difference between the dilatation of Air, of its own
accord, and by conftraint. For while it is willingly ex-
panded,

panded, the *Benſil* begins to grow ſlack, and remiſs, and
loſeth by degrees of its ſtrength; even as the *Spring* of a
Watch by the motion of the Wheels, becomes remiſs. But
when the dilatation is made by heat, and the Air compelled
to expand and open it ſelf, the *Benſil* becomes the ſtronger,
and the Preſſure the greater. Notwithſtanding, though
the *Benſil* of this incloſed Air G F, may be made ſtronger
by heat, to the expulſion of the Water F D, yet if this
rarefaction continue any time, the *Benſil* becomes dull and
ſlack. And the reaſon is, becauſe Air cannot be expan-
ded and opened to any quantity; an inch cannot be dilated
and opened to an hundred, or to a thouſand : neither can
the *Benſil* of it be intended, and increaſe to any degree, *v.g.*
from one to 20, 30, or 100. And therefore, as the ex-
panſion grows, the *Benſil* muſt at length ſlacken. But if
ſo be the Air were incloſed, as in a bladder knit about the
neck with a ſtring, then the more heat, the more *Benſil*:
for in this caſe there is a growth of Preſſure, without dilatation. And ſometimes the *Benſil* may be ſo intended with
the heat, that the ſides of the bladder will burſt aſunder.

From this Experiment we ſee firſt a confirmation of the
21 Theorem, namely; that there may be as much *Benſil*
and Preſſure, in the ſmalleſt quantity of a Fluid, as in the
greateſt; as is clear from the *Benſil* of the Air G F, which
in effect counterpoiſeth the weight of the whole *Atmo-*
ſphere, reſting upon the ſurface of Water A B. We ſee
ſecondly, that when the *pondus*, and the *potentia* of two
Fluids, are in *equilibrio*, or of equal ſtrength, a very ſmall
addition to either of them, will caſt the ballance. For if a
man ſhould but breath ſoftly upon the ſide of the Glaſs between G and F, or lay his warm hand to it, the ſaid Air
will preſently dilate it ſelf, and by becoming thus ſtronger,
thruſt

thruſt down the Water, and ſo overcome the *potentia* of
the ſurface. We ſee thirdly a confirmation of the ſixth
Theorem, namely, that the Preſſure of Fluids is on every
ſide; as is clear from the incloſed Air G F, that not only
preſſeth down the Water F D, but with as great force
preſſeth up the top of the Glaſs within, and preſſeth upon
all the ſides of it within, with the ſame force. This Experiment alſo, leads us to the knowledge of two things :
Firſt, of the reaſon, why with cold the Water aſcends in
the common *Weather-glaſſes* ; and why in hot weather
the Water deſcends. Secondly, from this Experiment
we may learn to know, when the Air is under a greater
Preſſure, and when under a leſſer : becauſe when the Air
becomes heavier, as in fair weather, the Water creeps
up in ſome meaſure, it may be two or three inches ; when
there is no alteration as to heat and cold : and in foul weather, or in great winds, when the Air is really lighter, the
ſaid Water creeps down as much. If it be asked, how ſhall
I know, whether it be the cold of the Air, or heavineſs
of the Air, that cauſeth the Water to aſcend ; and whether it be the heat of the Air, or the lightneſs of the Air,
that cauſeth the Water to deſcend ? I have propoſed this
queſtion of purpoſe, to let you ſee a miſtake. Many believe, that the aſcent and deſcent of Water in common
Weather-glaſſes, is allanerly from the heat and coldneſs of
the Air ; and therefore they conclude a cold day to be, becauſe the Water is far up : whereas the Water hath aſcended ſince the laſt night, by reaſon of a greater weight in
the Air, which alwayes is, when the weather is dry, and
calm, though there hath been no alteration of heat to cold.
If it be asked, how come we to the knowledge of this, that
the preſſure and weight of the Element of Air, is ſometimes

times less, and sometimes more? I answer, this secret o
Nature, was never discovered, till the invention of the
Torricellian Experiment, otherwise called the *Baroscope*.
For after the falling down of the *Quick-silver* to 29
inches; if you suffer it to stand thus in your Parlour or
Chamber, according as the Pressure, and weight of the
Element of Air, becomes more or less, so will the Altitude of the Mercury become less or more, and vary sometimes above 29 inches, and sometimes below. This alteration is very sensible, which is sometimes the tenth part
of an inch, sometimes the sixth, and sometimes the third,
according as the weight of the Air is less or more. From
December to *February*, I found the alteration become
less and more from 30 inches to 28, which will be three
fingers breadth. The common *Weather-glasses* then are
fallacious, and deceitful, unless they be so contrived, that
the Pressure of the Air cannot affect them, which is easily
done by sealing them *Hermetically*, and in stead of common
Water, to put in *Spiritus Vini rectificatissimus*, or the most
excellent Spirit of Wine, and strongest that can be made.

It may be here inquired, whether or not, Mercury would
ascend in this Glass, as the Water does? I answer it would;
because the ascent depends only upon the Pressure of the
Air, incumbing upon the stagnant Liquor in the Vessell,
that's able to drive up Mercury as well as Water. It may
be inquired secondly, how far Mercury will ascend, and
how far Water will creep up? I answer, Mercury can ascend no higher in a Tub, than 29 inches; and Water no
higher, than 34 foot; and this onely happens, when there
is no Air above the tops of the Cylinders to hinder their
ascents. But when there is Air, as G F above the liquor,
it can go no higher, than the point to which the cold is
                                                        able

able to contract the inclosed Air, which is in this Glass, the point F. It may be inquired thirdly, which is the greater difficulty, whether or not Mercury, will rise as easily in a Tub as Water; for seeing, its 14 times heavier, it seemes the Air should have greater difficulty to press it up, than to press up Water? I answer, 'tis greater difficulty for the Air to press up 20 inches of Mercury, than to press up 20 inches of Water; yet its no greater difficulty, for the Air to press up 20 inches of Mercury, than to press up 23 foot of Water, because the burden and weight is the same. It may be inquired fourthly, whether or not, it be as easie for the Air, to press up a thick and gross Cylinder of Water, as to press up a thin and slender one? For example, whether is it as easie for the Air to press up a Cylinder of Water 10 inches in Diameter, and 10 foot high, as it is to press up one, two inches in diameter, and 10 foot high? I answer, there is no more difficulty in the one, than in the other: and the reason is, because Fluid bodies do not counterpoise one another according to their *thickness*, but only according to their *altitude*, according to the fourth Theorem. Therefore seeing the slender Cylinder is as high as the grosser, it must be no more difficult to the Air, to press up the one then the other.

There is one difficulty yet remaining, which is truely the greatest of all; namely what's the reason, why its more difficult to the Air, to press up 20 inches of Mercury, than to press up 20 inches of Water: or more difficult to the Air, to press up 20 inches of Mercury, than to press up 10? I answer, this comes to pass, because the Air is more burthened with 20 inches of Mercury, than with 10. Now, if this be, then surely it must be more hard to the Air, to do the one, than to do the other: even as it is more

hard;

hard; for a man, to lift up from the ground, 20 pound of
iron, than to lift up 10 or 15. The cafe may be better il-
luftrated after this manner. Suppofe a man ftanding on the
ground, with a rope in his hand, coming down from a Pul-
ley above, drawing up a weight to the top of the houfe:
put the cafe likewife, the weight be a ftone of 20 pound,
and the weight of it, to increafe fucceffively, as it is pulled
up. Now its eafie for the man to pull up the ftone the
firft fathom; becaufe it is but 20 pound weight: but the
ftone becoming 40 pound in the fecond fathom, and 60
in the third, and 80 in the fourth and fo forth, untill it become 1000, he will find the greater difficulty, the longer
he pulls. 'Tis juft fo with Air, or Water, raifing Mercury in a Tub; for as the Cylinder of the Mercury grows
higher by rifing, fo it becomes heavier, and confequently
the imaginary furface, upon which the *Bafe* of the Pillar
refts, is more and more burdened, and fo becomes lefs and lefs
able to prefs it up. This leads us to a clear difcovery of the
reafon, why 'tis more difficult by fuction, to pull up Mercury in a Pipe, than to pull up Water; and more hard to
fuck up ten foot of Water, then to fuck up five. For trial
of this, which is foon done, take a flender Glafs-pipe 30
or 40 inches long, open at both ends, and drown the one
end among *Quick-filver,* and put your mouth to the other,
and having fucked, you will find greater difficulty to pull
up thorow the Pipe 15 inches of Mercury, than to pull up
10, or 8; and far greater difficulty to fuck up 20, than to
pull up 15. It may be objected, that if a man had ftrength
fufficient in his Lungs, to fuck out the whole Air of the
Pipe, thirty inches of Mercury would come as eafily up, as
three, which feemes to prove, that the difficulty of the
Mercurie's up-coming, depends not upon the weaknefs of

I

the

the Air, but upon the weakneſs of the Lungs, and want of
ſtrength to ſuck. I anſwer, though a man were able to
ſuck out the whole Air of the Pipe, yet 30 inches, will never aſcend ſo eaſily, as ten, nor ten ſo eaſily as three; and that
for the reaſons already given. But why is it then, (ſay
you) that the ſtronger the ſuction be, the higher the
Mercury aſcends in the Pipe? I anſwer, the ſuction ſerves
for no uſe, but to remove the impediment, that hinders the
Mercury from coming up, which is nothing elſe, but the
Air within the Pipe. Now, the more of this Air that's
taken away by ſuction, (the ſtronger the ſuction is, the
more Air is taken away) the farder up comes the Mercury.
But why ought there to be difficulty in the ſuction of Mercury, to the altitude of 15 or 20 inches, more than in the
ſuction of Water to that altitude? I anſwer, when I ſuck
Water up thorow a Pipe, the ſuction of the Air above it,
is eaſie; becauſe the aſcending Water helpes much to drive
it up to the mouth, the outward Air driving up both.
But the ſuction is difficult in Mercury, becauſe the aſcending liquor, does not help ſo much, to drive up the Air to
the mouth, as the Water does. And the reaſon is, becauſe the Air, being more burdened with 15 inches of Mercury, than with 15 inches of Water, cannot ſo eaſily drive
up the one as the other, and ſo Mercury cannot ſo eaſily
drive up the Air of the Pipe to the mouth, as Water does.
In a word, according to the difference of *ſpecifick* weight,
between Water and Mercury, ſo is the difficulty of ſuction; therefore, becauſe Mercury is 14 times heavier than
Water, there is 14 times more difficulty, to pull up the one,
than the other. Note, that *ſuction* is not taken here
ſtrictly, as contradiſtinguiſhed from *pulſion*; but in a large
ſenſe, as it may comprehend it.

To

To proceed a little further, let us suppose the Pillar of Mercury (see the 11. Figure) G H, that's raised by the surface of Air F G, to be 29 inches, and every inch to weigh one ounce. Secondly, that the said surface has 29 degrees of power or force in it: for in all counterpoises the *Pondus* and the *Potentia* are equal; therefore, if the Mercury be 29 inches, the *Potentia* of the surface must have 29 degrees of strength or force in it, to counterballance the *Pondus*. These things being supposed, which are evident, let us imagine the surface of Air, to raise the Mercury one inch above F G. In this case, the surface is weaker than it was; which I prove evidently, because it is now but able to raise 28 of Mercury. Imagine next, the said surface to have raised the Mercury two inches above F G, then it follows, that it must be yet weaker, because it's now but able to raise 27 inches: for by supporting two ounce of the *Pondus*, it loseth two degrees of it's own *Potentia*. In raising three inches of Mercury, it is three degrees weaker; and in raising four, it is four degrees weaker, and so forth; therefore, having raised 28 inches, there is but one degree of force remaining in the surface. And when it hath raised the whole, namely 29, it is no more able, and can no more press. For confirmation, put the case that the surface of Air F G, were as able, and had as much Pressure in it, after it hath raised 29 inches of Mercury, as it is after the raising of 10; then it follows of necessity, that after the raising of 20, it shall raise 19 moe, which is impossible, seing the greatest altitude is 29. It follows of necessity, (I say) because after the raising of 10, it is able to raise 19 moe: therefore if it be as able after 20, as after 10, it must raise 19 after 20. Yea, if it be as able after 20 as 10, it must be as

able

able after 29 as 10. If this be, then it may raise other 29,
and a third 29, and so *in infinitum*. Therefore, I conclude, that when two Fluid Bodies are in *equilibrio* one
with another; or when the *pondus* is equal to the *potentia*, none of them doth actually press upon another, at least
the surface hath lost all its Power and Pressure, which is
also evident in the Pillar. For understanding this, let us
suppose A C B (Figure 11) to be a Pipe 58 inches long,
and full of Mercury, and every inch of it to weigh one
ounce. Now, when the orifice D is opened, there is here
as great an inequality, between the *pondus* and the *potentia*
of the surface of Air E B, on which it rests, as was between the surface F G, and the *pondus* of Mercury H G.
For as F G had 29 degrees of power to raise G H, so the
Pillar A B has 29 ounce of weight, to overcome the surface E B. And as the surface F G, became one degree
weaker, by raising one inch of the Mercury H G, and two
degrees weaker, by raising two inches, and so forward, till
it lost all its Pressure; so the Pillar, by falling down one
inch, loseth one ounce of the weight; by falling down
two, it loseth two ounce, and so forward, till by falling
down from A to C, it loseth all its Weight and Pressure.

But here occurreth a difficulty; for if the surface F G,
hath lost all its Pressure, by raising the Mercury from G to
H; and if the Pillar C B, hath lost all its Pressure, by falling down from A to C; it follows, that when a Pillar of
a Fluid, and a surface of a Fluid are in equal termes, or
brought to an *equipondium*, there is no Pressure in them at
all. For answer, consider first, that in all counterpoises,
there are necessarily two things, the *movens* and the *motum*; the thing that moves, and the thing that is moved.
Secondly, you must consider the *motum*, to have a *pondus*

or

or weight in it, and the *movens* to have a *potentia*, or power, wherewith it moves that weight. Thirdly, that as the thing that moves, hath a power or force in it ſelf, whereby it moves, ſo the thing that is moved hath a power or force in it ſelf, whereby it reſiſts the motion, Fourthly, that ſometimes the reſiſtance of the thing *moved*, may exceed the power of the *movent*, as when a Quarrier with a Leaver, endeavours to prize up a ſtone too heavy for him : or the power of the *movent*, may exceed the reſiſtance of the *weight*; or both may be of equal power. Conſider filthly, that as the *pondus* of the thing moved, begins to grow more and more, ſo the power of the *movent* decreaſeth proportionably ; not abſolutely , as heat is extinguiſhed in Water by the cold Air, when it is removed from the Fire, but reſpectively. For example, when a man holds a ballance in his hand, with ſix pound in the one ſcale, and but one pound in the other, if you add another pound, the weight grows more, and the power and force of the oppoſite ſcale grows leſs proportionably ; not *abſolutely*, for it ſtill remains ſix pound, but *reſpectively* : that's to ſay , ſix pound is leſs in reſpect of four, than in reſpect of five; or the reſiſtance of ſix pound is leſs, two counterpoiſing it, than being counterpoiſed by one. When a third is added, the weight grows yet more, and conſequently the reſiſtance of the oppoſite ſcale becomes yet leſs, till by adding the ſixth and laſt pound, you augment and encreaſe the *pondus* to that ſame degree of ſtrength, that the reſiſtance of the oppoſite ſcale is of. From theſe conſiderations, I ſay, the ſurface of Air F G, hath not loſt all its Preſſure *abſolutely*, by raiſing the Mercury from G to H, but only *reſpectively*, becauſe it ſtill retains 29 degrees of force in it ſelf. I ſay *reſpectively*, becauſe when the Mercury is
raiſed

raiſed ten inches, the power of the Air which is of 29 degrees of force, is leſs in reſpeſt of ten ounce, then in reſpeſt
of five; or the power of 29 degrees of force is leſs, being
counterpoiſed by ten ounce, than being counterpoiſed only by five. And when it is raiſed 20, it is yet leſs in this
reſpeſt, than in reſpeſt of ten. And when it has raiſed
the Mercury to the greateſt altitude H, it may be ſaid to
have loſt all its Preſſure, ſeing it is not able, by vertue of
a counterpoiſe, to do any more. Even as ſix pound in this
ſcale, may be ſaid to have loſt all its reſiſtance and weight,
by putting in the other ſcale, firſt one pound, next two
pound, and then three pound, till the laſt be put in, at
which time it hath no more reſiſtance. Though this be,
yet it ſtill remains ſix pound. Even ſo, the Air F G ſtill
remains of the ſame force and power, while it ſuſpends
the Mercury G H, that it was of before. Likewiſe, the
Pillar A B, cannot be ſaid to have loſt all its preſſure *abſolutely*, by falling down from A to C, but only *reſpectively*,
becauſe the ſaid Pillar C B, is ſtill 29 ounce weight. I
ſay *reſpectively*, becauſe in falling down ten inches, or in
loſing ten ounce, the weight that's now but 48, is leſs, in
reſpeſt of 29, than while it was 58. It is yet leſs, when it
hath fallen down other ten, becauſe being now but 38, it
muſt be yet leſs in reſpeſt of 29, than 48. And when it
hath fallen down to C 29, it may be ſaid to have loſt all its
weight, becauſe it can do no more, having *reſpectively* loſt
all its Preſſure.

From what is ſaid, we ſee a clear ground to diſtinguiſh
in Fluids a *pondus* and a *potentia*. Secondly, that the *potentia* may ſometimes exceed the *pondus*, and contrariwiſe
the *pondus* may exceed the *potentia*. Thirdly, that inequality of weight, between the *pondus* and the *potentia*, is
the

the cause of motion of Fluids. Fourthly, that the motion never ceaseth, till the *pondus* and the *potentia* become of equal force. This conclusion is not so universal as the rest, because the motion may sometimes cease, before this be. For example, when the Air is pressing Mercury up thorow a Tub shorter then 29 inches, the motion ends before there be a perfect counterpoise; for 20 or 15 inches of Mercury, can never counterballance the force and power of the Air. In such a case then, there is an unequal Pressure, the Air pressing the Mercury more, than the Mercury doth the Air.

EXPERIMENT VIII.
Figure 12.

TAke the Vessel A B C D, and fill it with Water, as high as H I. Take next a Cylinder of *stone* F G, and drowning the half of it among the Water, suspend it with a chord to the beam N O, with a ring at E. Now in this case, though the stone do not touch the bottom of the Vessel, yet the Water becomes heavier, than before. For discovering the true reason of this, I suppose first, the weight of the Water, before the *stone* be drowned, to be 40 pound. I suppose next, that after the *stone* is drowned, the said Water to weigh 50 pound. And lastly, the *stone* to weigh 60 pound. I say then, the Water must be 10 pound heavier than before, because it supports 10 pound of the *stone*. 'Tis certain the beam is less burdened by 10 pound than before. If this be, then surely the Water must sustain it. It were great temerity and rashness, to averr that neither the Beam, nor the Water sustains it, which

which is really to say, it is sustained by nothing. It cannot be said without ignorance, that 10 pound of the *stone* is evanished, and turned into a *Chimera*. If it be said, how can such a Fluid Body as Water, be able to support any part of the weight of the *stone*, that is such a heavy Body? I answer, there is here no difficulty, for if the imaginary surface K L, upon which the 10 pound of the stone rests, be able to sustain 10 pound of Water (I suppose the *stone* taken away, and the place of it filled with Water) then surely it must also be able to sustain 10 pound of the heaviest metal; seing ten pound of Lead, or Gold, or Stone, is no heavier than 10 pound of Water. If some say, this rather seems to be the reason, why the Water becomes heavier, after the *stone* is drowned, because it possesseth the place of as much Water, as would weigh 10 pound; not (as was said) because the Water supports 10 pound of it. Therefore it may be judged, and thought, that if the space that the *stone* occupies, were filled with Air, or some light Body, without sensible weight, the Water would become heavier than before. For example, if in stead of the *stone*, there were placed a *bladder* full of wind, within the Water, and tied to the bottom with a string, that the surface might swell from H I to A B, the Water of the Vessel would become as much heavier than before, as is the bulk of Water, equal to the quantity of the *bladder*. Therefore, the Water becomes heavier, not because it supports any part of the *stone*, but because the *stone* occupies as much room and space, as would contain 10 pound of Water: for by this means the drowned *stone* raiseth the Water from H I to A B; and so the Cylinders A C, and B D, being higher, press with greater weight upon the bottom C D, even with as much more weight, as if the space that the *stone* occupies were filled with VVater.

For answer to this, we shall make this following Experiment. Take the Vessel M P V X, and fill it with VVater to Q R. Next, take a large *bladder* W Y full of wind, and tying the neck with a threed, thrust it below the Water, and fasten it to the bottom, with a string, to the Ring Z. This done, the Water swells, and rises from Q R, to M P. Now, if it be true, that the Water in the Vessel becomes heavier, not because it supports 10 pound weight of the *stone*, but because the *stone* occupies the room of 10 pound of Water; then it ought to follow, that after the *bladder* is tyed below the Water, the said Water should become heavier, than before, even by three pound; for I suppose a bulk of Water, equal to the bulk of the *bladder*, to weigh as much. And the reason is, because (as you say) the quantity of the *bladder* W Y, makes the water swell from Q R to M P, by which means the Pillars of Water M V, and P X becomes higher, and so presseth with greater weight upon the bottom V X. For clearing this difficulty, I say, when a *bladder* is thus below the VVater, tyed to the bottom, the VVater becomes not three pound heavier: for when you place the Vessel with the VVater and *bladder*, in the Scale of a Ballance, the said VVater weighs no more, than if it wanted the *bladder*: therefore the VVater becomes not heavier, because the *stone* possesseth the room of 10 pound of Water, but because the Water sustains 10 pound of the *stone*. Now the reason, why the *bladder* makes not the water heavier, though it raise it from Q R to M P, is this; because though verily there be a greater Pressure then before, even upon the bottom of the Vessel, yet because moe parts are not added,

K   the

the *natural weight* cannot be augmented, which essentially depends upon the addition of these parts, If it be replyed, the Experiment of the *bladder* is to no purpose, because it being knit to the bottom, pulls up the Vessel, with as great force, as the growth of the Pressure bears it down, and so the *Bladder* cannot make the Water heavier. But, if so be, it were possible, that the *Bladder* could remaine within the middle of the Water, without being knit to the bottom, and consequently without pulling up the Vessel, then surely the Pillars of Water M V, and P X, being higher, would press with greater weight upon the bottom, and so make the Vessel, and the Water weigh more in the ballance: for 'tis to be supposed, that during all this time, this Vessel with the Water, is in one scale, and a great weight of stone or lead, in the other. So would the Water A B C D become heavier likewise, provided the space and room, that the *stone* fills among the Water, remained intire, after the stone is taken away: because that room and empty space remaining, would keep the surface, as high as A B, by which means, the Pillars A C and B D, being higher, would press with greater weight upon the bottom, and cause the Water weigh more in the ballance. I answer, though by some extraordinary *power*, the *bladder* could remain below the water, of its own accord, as it were, and though the space and room, by that same *power*, which is left by the *stone*, were keeped empty, yet shall they never be able to make the Water heavier. As to the reason, that's brought, I answer, the rising and swelling of the Pillars, will make indeed a greater Pressure upon the bottom of the Vessel, but because this Pressure may be produced, and generated without the addition of new parts, therefore, it can never make the Water heavier: for if this

were

were true, then it would follow, that the more a body is comprest, it should be the heavier, which is contrary to sense, and experience. This Pressure is like unto *Bensil*, that cannot weigh in a ballance, though the *thing bended* do weigh; as a Bow that weighs so many pounds, but the *Bensil* of it weighs nothing : Next, will any man think, that a Cub of Water six foot high, and six foot thick, will weigh more in a ballance, then it did, after it is turned into a long square Pillar 216 inches high? I grant, there is near 60 times a greater Pressure, upon the bottom of the Vessel, yet because this Pressure is generated, without the addition of new parts, it cannot make the Water heavier. Moreover, it is *mechanically* possible to keep the Water S T V X, under that same degree of Pressure it hath, though the rest above were taken away : if this be, then it ought to be as heavy, as the whole, seing it still Presses the bottom, with that same degree of Pressure, it had from the whole: but what is more absurd, than to say, one part of Water, is as heavy, as the whole? *e. g.* a pint as heavy as a gallon. If it be said, the Pressure, and the weight, are but one thing, at least *effectively*, which is sufficient to the purpose in hand, as is clear from the Theorem 23. I answer, they are but one thing indeed, in order to the *Ballance of Nature*, but they are neither *formally*, nor *effectively* the same thing in order to the *Libra* or *Artificial Ballance*, whereof we are now treating. I shall conclude with this; while the Vessel with the Water, is thus placed in the Scale of the Ballance, and in *equilibrio*, with the opposite Scale, cut the string that tyes the *bladder* to the bottom, and when it comes above, you will find the Water, just of the same weight it was of: for though the surface M P, by taking out the *bladder*, settle down to Q R, yet there's no alteration made in the weight. From this I gather, that if the swelling of the VVater should make it heavier, then the subsiding and falling down of it, ought to make it lighter.

From these Experiments we gather first, that in VVater there is a Pressure, because it sustains 10 pound of the *stone* FG. Secondly, that whatever heavy body is weighed in Water, it loseth just as much of its weight, as the bulk of Water weighs, it puts out of its place. This is evident, because the *stone* is 10 pound lighter in VVater, than in the Air, because the VVater that would fill the room of the *stone*, is just of that weight. VVe see thirdly, that the *Pressure* of VVater, and the *natural weight* of it, are two things really distinct; because the *Pressure* may be augmented, without any increment of the *natural weight*. VVe see fourthly, that the *Pressure*, or *Bensil* of a Fluid, cannot affect the Scale of a Ballance, but only the *natural weight*. VVe see fifthly, that a body naturally heavier than Water, *weighs in Water*, because the *stone* FG, makes the Water about it, 10 pound heavier. If it be inquired, whether bodies, that are naturally lighter, will *weigh in Water* ? I answer, if they be of any sensible weight, they weigh, as well as the other. For this cause, I except Air. For though they were never so light, in respect of Water, yet if they have any considerable gravity with them, they will make the Water heavier, they are among. Put the case the Body were a Cube of Timber of six inches, weighing sixteen ounces, and that a Cube of Water of that quantity, weighed 112 ounces. Here's a great inequality, between their *natural weights* : yet if that piece of Timber, were made to exist in the middle of Water, as the *Bladder* doth, it would make it 16 ounces heavier.

The

The reason is this; these 16 ounces are either supported by a surface of Water, or they support themselves. This last is impossible. If the VVater support them, then must they make the said VVater 16 ounces heavier. Note, that though a Body naturally lighter then VVater, as *Cork*, may be said to *weigh in Water*, that's to say, to make it heavier, in which sense VVater *weighs in Water*, because if you add a pint to a gallon, it makes it heavier; yet if you take a piece of *Cork*, and knit it to the Scale of a Ballance, by a threed, the *Cork* hanging among the VVater, the Scale hanging above in the Air, it will not *weigh in Water*; because in this sense, no Body *weighs in Water*, but that which is naturally heavier then VVater, as Lead, or Stone. In this sense, VVater doth not *weigh in Water*, as will be seen in the 17 Experiment.

EXPERIMENT IX.
Figure 13.

Take a Glass-pipe 70 inches long or there-about, and of any wideness, having the upper end H, *hermetically* sealed, the *lower* end C compleatly open, and fill it with Mercury, and cause a *Diver* carry it down to the ground of the sea M N, where I suppose is standing the Vessel A B D E with stagnant Mercury, and drown the end below the surface A B. This being done, the Mercury falls from the upper end H, to the point G, and there halts; the space H G being empty. For understanding this Experiment, I shall propose several questions, and answere them. First, what's the reason, why the Mercury subsides, and sinks down from H to G? I answer, as formerly

merly

merly in the like cases, inequality of weight between the *Pondus* of the impending *Quick-silver*, and the *Potentia* of the surface, of the stagnant *Quick-silver* D C E. For while the Tub is compleatly full, the weight is so great, that the surface D C E, is not able to sustain it, therefore it must fall down, seing motion necessarily followes in Fluids, upon inequality of weight. It may be inquired secondly, why it halts at G, 58 inches from A B, and comes no further down? I answer it halts at G, because when it hath fallen down to that point, there happens equality of weight, between the suspended Pillar, and the foresaid surface: for whatever weight the said Pillar is of, the surface on which it rests, is of the same. In a word, the *Pondus* of the one, and the *Potentia* of the other are now equal. For understanding this, consider according to the 25 Theorem, that the weight of the Element of Air, upon the surfaces of waters, is equivalent to the burden of 34 foot of water, therefore the first and visible surface of this Water L I K, is really as much prest, with the burden of the *Atmosphere*, as if it had 34 foot of Water upon it. Consider next, that between the said surface, and the ground M N, are 34 foot of Water indeed. Consider thirdly, that a Pillar of Water 34 foot high, is exactly of the same weight, with a Pillar of Mercury 29 inches high; for if Water be 14 times lighter than Mercury, then they cannot be of equal weight, unless the one be 14 times higher than the other. Now, supposing the weight of the Air upon the surface L I K, to be equivalent to 34 foot of Water, or (which is the same thing) to 29 inches of Mercury, the surface of the stagnant Mercury A B, must be as much burdened with the incumbing Water, and the Air together, as if it had really resting upon it, a Pillar of

Mercury

Mercury 58 inches high. If this be, then it follows by necessity, that there must be an equality of weight, between the *pondus* of the Mercury in the Tub, and the *potentia* of the surface D C E; Or (which is all one thing) that the part C, on which the Pillar rests, is no more burdened, than the part D or E. For if 34 foot of Water, and 34 foot of VVater, be equivalent for weight, to 58 inches of Mercury, then must the part D and E, be as much burdened with the said weight, as the part C is burdened with the Pillar within the Tub, seing both are of the same height: therefore the power, and force of the imaginary surface of the stagnant Mercury D C E, is of the same strength, with the weight of the Pillar G F B. And this lets us see the reason, why the whole 70 inches cannot be suspended; for if the outward Pressure that's upon A B, be but equivalent to the Pressure of 58, it can never make the surface D C E able to support 70.

To make it evident (if any doubt) that the Mercury is suspended by the weight of the Water, and the weight of the Air superadded, let a *Diver* bring up this Engine to the top of the Water, and he will find the one half to have fallen down, namely from G to F, the other half F B remaining. And if it were possible, to convey this Experiment to the top of the Air, the Bearer would see, the remaining half to fall down likewise, and become level with A B; for where no Pressure of Air is, there can be no Mercury suspended. This falling down, is not all at once, but by degrees, and keeps a proportion with the Pressure of the Air, that grows less and less, from the ground to the top.

From this Experiment we see first, the great Pressure and weight, the Elements of Air and Water are under,

seing

feing this Water, that's but 34 foot deep, fuftains the Mercury between G and F, 29 inches, as much between F and E, being kept up by the Preffure of the Air. We fee fecondly, that this Preffure is according to *Arithmetical Progreffion*, as 1, 2, 3, 4, 5. becaufe in going down the firft 14 inches, the Mercury rifes one inch; in going down the fecond 14 inches, it rifes two; in going down the third 14 inches, it rifes three, and fo forward. We fee thirdly, though a VVater were 100 fathom deep, yea 1000, yet the Preffure of the Air above is found at the bottom: for fuppofing this Experiment were 100 fathom deep, yet would the Air from above have influence upon it, to fuftain fo many inches of the Mercurial Cylinder. A *Diver* then, 10 or 15 fathom under the VVater, muft be burdened with the weight of the Air, as well as with the weight of the VVater, fo muft the Fifhes, though never fo deep. We fee fourthly, that the parts of a Fluid cannot ceafe from motion, fo long as there is an inequality of weight between the *pondus* and the *potentia*. This is clear from the falling down of the Mercury from H to G. And affoon as equality of weight happens, the motion ends. This is clear from the Mercurie's halting at G. Fifthly, that in Mercury, as well as in Water, or Air, furfaces may be diftinguifhed, and that thefe furfaces, are endowed with a *Potentia* or power, begotten in them by fuperior and extrinfick weight. This is clear from the imaginary furface D C E, that's made powerful to fupport 58 inches of Mercury in the Tub, and that by the weight and Preffure of the Air refting upon A B. Sixthly, that, as two Fluids differ in *fpecifick* and natural weight, fo they differ in *altitude*, when they counterpoife one another. This is clear from the difproportion that's between the altitude

titude of the Mercury suspended, and the height of the Water, and Air suspending. G F then is 29 inches, and the deepness of the Water from K to N is 34 foot, because Water is naturally 14 times lighter than Mercury. F B is likewise 29 inches, and the hight of the Air, that rests upon the surface of Water is six or seven thousand fathom high; because Air is 14000 times *naturally* lighter than Mercury. Seventhly, that Fluid Bodies counterpoise one another, not according to their *thickness* and *breadth*, but only according to their *altitude*. This is evident; for though this Tub were never so wide or narrow, yet the altitude of the Mercury is unchangeable. Hence it is, that the *thickest* Pillar of Water in the Ocean, is not able to suspend more Mercury, than the *slenderest*, I mean as to altitude. And hence it is, that the smallest Cylinder of Mercury, no thicker than a silk threed, is able to counterpoise a Pillar of Water, of any thickness whatsoever. We may conclude lastly, that when a *Diver* is 20 fathom under the Water, he is under as much burden, as if he were under 14 or 15 foot of Quick-silver. Suppose a man lying on his belly, within a large Vessel, and 14 or 15 foot of Mercury poured in upon him, surely it may be thought, that such a burden were insupportable. But put the case, the *Diver* were down 40 fathom, then must the burden be doubled. This follows, because if a Pillar of Water 34 foot high, with the weight of the Air superadded, be as heavy, as 58 inches of Mercury, then surely a Pillar 20 fathom high, or 100 foot, must be as heavy as 170 inches, which is more than 14 foot.

5 foot = 1 fathom in this Aut

L   EXPE-

EXPERIMENT X.
Figure 14.

AGainſt the former Experiment, there occurres ſome difficulties, which muſt be anſwered. As firſt, if it be the Preſſure of the Water, that ſuſtains the Mercury in the Tub (ſee the 13. Figure) then the weight of the ſaid Mercury ought not to be found, while the Tub is poiſed between a mans Fingers. But ſo it is, that when a *Diver* grips the Tub about the middle, and raiſes it a little from the bottom of the Veſſel, he not only finds the weight of the Tub it ſelf, but the weight alſo of the 58 inches of Mercury that's within it. But this ought not to be, if the ſaid Mercury, be ſuſtained by the outward Water. In a word, it ought not to be found, becauſe the ſaid Pillar of Mercury, as really ſtands, and reſts upon the imaginary ſurface D C E, as a Cylinder of Braſs or Stone, reſts upon a plain Table of Timber or Stone. If then, it be ſupported by the ſaid ſurface, why ought I to find the weight of it, when I lift up the Pipe a little from the bottom of the Veſſel ? For clearing this difficulty, conſider, that when the Mercury falls down from H to G, it leaves a ſort of *vacuity* behind it, wherein there is neither Air nor Water. Conſider ſecondly, that for this cauſe, there happens an unequal Preſſure ; the top of the Tub without, being burdened with the Pillar of Water I H, which actually preſſeth it down, and nothing within between G and H, that may counterballance that downward Preſſure. Theſe things being conſidered, I anſwer to the difficulty and ſay, it is not the weight of the ſuſpended Mercury that I find, but the weight of the Pillar of Water I H, that reſts up-

on

on the top of the Tub. If it be ſaid, the Preſſure of a
Fluid is *inſenſible*, and cannot be found. I anſwer, it's true,
when the Preſſure is equal and uniform, but not when the
Preſſure is unequal, as here. If it be asked, how comes
it to paſs, that the Pillar of Water I H, is exactly the
weight of the 58 inches of Mercury? I anſwer, beſides
the ſaid Pillar, there is another of Air, that reſts upon the
top of it, which two together are exactly the weight of the
ſuſpended Mercury; I H being of the ſame weight with
the Mercury G F, and the foreſaid Pillar of Air, being of
the ſame weight with the Mercury F B. To make it more
evident, remember that one inch of Mercury, is exactly
the weight of 14 inches of Water; and that one inch of
Mercury, is of the ſame weight with 14000 inches of Air.
If this be, then muſt the Pillar of VVater I H, that's 34
foot high, and of the ſame thickneſs with the 29 inches of
Mercury G F, be of the ſame weight with it, ſeing 29
inches are to be found 14 times in 34 foot. For the ſame
reaſon, is the Pillar of Air, namely S I, that reſts upon
the top of the Pillar of VVater I H, of the ſame weight
with the 29 inches of Mercury F B. For after a juſt reckoning, you will find, that 29 inches will be found 14000
times in the Pillar of Air, that reſts upon the Pillar I H.
Or in a word, the hight of the Air is 14000 times, 29
inches.

But here occurrs another difficulty. Let us ſuppoſe there
were a Tub ſix foot high, one inch wide, having the ſides,
3 inches thick. Imagine likewiſe the ſaid Tub to be under the water 34 foot, with 58 inches of Mercury in it, as
is repreſented in this 14 Figure. This being ſuppoſed,
the Pillar of Water E A F C G D, muſt be far heavier,
than the 58 inches of Mercury H B. The reaſon is clear,

becauſe

becauſe the ſaid Pillar , is not only 34 foot high, but as thick, as the Diameter of the Tub , whoſe ſides are three inches thick. I anſwer, the whole weight of that Water E A F C G D is not found, while a man poiſes the Tub between his fingers, but only the weight of the part G A, which is exactly the weight of the Mercury H B. But here occurrs the great queſtion, namely, why I find only the weight of the Water G A, and nothing of the weight of the Water, C E, or D F? I anſwer, I cannot find the Preſſure of the Water C E, becauſe it is counterpoiſed with the upward Preſſure of the Water I K. And for the ſame reaſon, I cannot find the weight of the Water D F, becauſe it is counterpoiſed by L M ; but becauſe there is nothing between H and A, to counterpoiſe the downward Preſſure of the Water G A, therefore I find that. If it be objected, that the Water I K, cannot counterpoiſe the Water C E, becauſe the one is ſarder down than the other, and conſequently under a greater Preſſure, than the other. I anſwer, though I K be ſtronger than C E, yet a compenſation is made by the weight of the Tub. For underſtanding this, let us ſuppoſe the Water C E, and D F, to preſs downward with the weight of ſix pound, and the Water K I, and L M, to preſs upward with the weight of ten pound, there being four pound in difference. Suppoſe next, the Tub to weigh in the Air ten pound, and in the Water only ſix pound. If this be, then according to the eighth Experiment, and eighteenth Theorem, four pound weight of the Tub muſt reſt upon the ſurface I L. And if this be , then muſt the Water I K, and L M, be four pound weaker with the Tub, than without it, and muſt only have ſix pound of upward Preſſure.

From

From these Experiments we conclude first, the truth of the tenth Theorem, namely that the weight of a Fluid is only found by force, when the Press is once uniform, and equal. This is evident from our finding the weight of the Pillar of Water IH, as in the 13 Figure. We conclude secondly, that in all Fluids, there is a *ponderosity* and a *protrusion*; as is clear from the *ponder* of Water EAFCGD, this presseth down the Tub, and the *protrusion* of the Water IKLM, that presseth up the same Tub. We see thirdly, that there cannot be two surfaces of Water differing in altitude, but they must differ in degrees of Pressure: because the surface EAF, a weaker, than the surface IL, that be not higher than this. We see fourthly, that two surfaces differing in strength, may be made equal by some Body or other intervening; because though IL be stronger than EAF, yet being it supports four pound of the Tub, is pressed up with no more force, than EAF, presseth down with. We see fifthly, that as a Body suspended in a Fluid, as in Air, or in Water, may have one part of its parts equally with that Fluid, and another part unequally: this is evident, because the parts E and F, are equally pressed with the Pillars CE, and DF, being this Pressure is counterpoised with the Pressure of Water IK, and LM. But the middle part of the Tub A, is unequally pressed, being it is pressed downward, with the Water GA, but not pressed upward with the Mercury BH. We see sixthly, that whatever be the thickness of a Pillar of a Fluid, yet no more of its weight is found, or is sensible, than the part, which presseth unequally: for though EAFCGD, be a Pillar six or seven inches thick, yet no more of the Pressure is sensible, than what comes from GA.

G A. VVe ſee ſeventhly, that a Body equally preſt with a Fluid, weighs leſs, but a Body unequally preſt, weighs none at all. This is clear in many particulars; for a Stone weighed in VVater, loſeth not all the weight, but a part, becauſe it is equally preſſed. But a Body unequally preſt, as is the Mercury H B, hath no weight at all, as it now ſtands. For underſtanding this, you muſt conſider, that the whole weight of it reſts upon the ſurface of VVater I L. Therefore though it could be weighed by a ſtring, paſſing from the top H, to a Ballance exiſting in the Air; yet the ſaid Ballance would find none of its weight, ſeing it is wholly ſuſpended by the VVater; but a Stone ſo weighed, is only ſuſpended in part, by the Water.

EXPERIMENT XI.

Figure 15.

A M Z C is a Water 15 foot deep. A B a Glaſs-tub 14 inches long, and full of Mercury. B C a Pillar of Water 13 foot, 10 inches high, thorow whoſe middle goes a ſtring to the ſcale of the Ballance K, exiſting in the Air. D E is a Tub full of Mercury 28 inches long, with a Pillar of Water above it E F, 12 foot and eight inches. G H a Tub 42 inches long, with a Pillar of Water above it H I, 11 foot and ſix inches high. And laſtly, A D G S M an imaginary ſurface, 15 foot deep. This Experiment is brought hither, to demonſtrate that a heavy Body, weighs as much in Water, as in Air, which is point-blank to the common received opinion, and deſtructive of the 18 Theorem. To evince this, I muſt ſuppoſe the 14

inches

inches of Mercury in the Tub A B to weigh 14 ounce;
and the 28 inches of Mercury D E , to weigh 28 ounce;
the 42 inches G H to weigh (I mean in the Air) 42 ounce.
Now I ſay , to make a juſt *æquipondium* between the two
Scales K and L, there muſt be 14 ounce put into the
Scale L. If after this manner you weigh the Tub and
Mercury D E , 28 ounces will be required in the Scale L,
and 42 , if you weigh the Tub and Mercury G H. For
proving this Doctrine, I muſt appeal to Experience, which
will not fail in this. If you reply, and ſay, upon ſuppoſition the Tub and Mercury G H, were a ſolid piece of braſs,
or iron thus ſuſpended in the Water, ought it not to weigh
leſs here than in the Air, even as much leſs, as is the weight
of the quantity of Water, it puts out of its place: why
then ſhould not the Pipe H G, with the Mercury in it, do
the ſame , ſeing there is no apparent difference between
them, as to this?

But to leave this, which will appear afterwards, and to
let the Reader ſee the truth of the 18 Theorem, I affirm,
'tis not the weight of the 14 ounces of Mercury A B, that
burdens the ſcale of the Ballance K, and that makes a counterpoiſe with the 14 ounces of Stone, or Lead, thats in the
ſcale L. What then is it , you ſay? I anſwer , 'tis 14
ounces of the Pillar of Water B C that does this. Neither doth the weight of the 28 ounces of Mercury D E
burden the Ballance, but only 28 ounces of the Water E F.
Neither doth the Ballance ſupport the weight of the 42
ounces of Mercury G H, but it is only burdened with 42
ounces of the Water H I. The reaſon is moſt evident,
becauſe according to the Principles of the *Hydroſtaticks*
already laid down, the Cylinder of Mercury A B, within
the Tub A B, reſts immediatly upon the imaginary ſurface

face of the Water A D G, and therefore cannot burden the scale in any wise. The same is true of the
other two Cylinders of Mercury. But in this I find small
difficulty. The greater is, how to make it out, that the
scale K, supports 14 ounces of the Water B C, and 28 of
the Water E F, and 42 of the Water H I. To make
this seem probable, consider first, as was noted, that this
VVater is 15 foot deep, and consequently the Pillar of
VVater B C, 13 foot 10 inches. The VVater E F 12
foot eight inches. And H I, 11 foot and a half. Consider secondly, though this be true, yet we must count the
Pillar of VVater Z M 49 foot high. The reason is evident, because the Pressure of the Air, upon the surface of
all Waters (according to the 25 Theorem) is equivalent
to 34 foot of Water: this then being added to 15, makes
49, and by this reckoning the Water B C is 47 foot ten
inches: the Water E F 46 foot eight inches: And lastly, the Water H I 45 foot six inches. Thirdly, for easie
counting, I must suppose the whole Cylinder Z M to
weigh 42 ounces, every 14 inches one ounce: and consequently the Water B C to weigh 41 ounces; the Water E F to weigh 40 ounces; the Water H I 39 ounces.
Note, that in *Physical* demonstrations, 'tis not needful to
use *Mathematical* strictness in counting; and so leaving
out fractions, we shall onely use round numbers. Consider fourthly, that in all Fluids, as hath been frequently
marked, there is a *pondus* and *potentia*, the Water B C being the *pondus*, and the Mercury A B the *potentia*, the one
striving to press down the Tub, the other striving to press
it up. Consider fifthly, that by how much the more a
Body suspended in a Fluid is pressed up, by so much the
less the weight that presseth it down is found: and contrariwise,

trariwife, by how much the lefs it is preffed up, by fo much
the more the Preffure above is found. Confider fixthly,
the lefs that a furface of Water is burdened, the more able
it is to counterballance the oppofite Preffure, and the more
it is burdened, it is the lefs able. Confider feventhly, that
the Mercury A B, (which is evident in all Fluids) not
only preffeth downward , and burdens the furface A D G,
but alfo preffeth upward, and therefore actually endeavours
to thruft up the Tub; and fo it is, that the Tub is preffed
between two, namely between the Water C B, and the
Mercury within it.

Now from thefe confiderations I fay, the fcale K, muft
fupport, and bear up 14 ounce of the Water B C : for feing
the Mercury is fupported by the furface of VVater on
which it refts, it cannot by any means burden the ballance
with its weight; and feing it actually preffeth up the
Tub , (according to the feventh confideration) it muft fo
much the more counterpoife (according to the fixth) the
oppofite Preffure of the VVater B C, and confequently
diminifh the weight of it : fo that the Ballance cannot fupport the whole, but a part. For according to what degrees of force, the Mercury preffeth up the Tub with, according to the fame, muft the Preffure upon the top of the
Tub be diminifhed, and fo if the Mercury prefs up the Tub
with the force of 27 ounce , the VVater B C muft prefs
it down with 14 ounce only, and fo the Cylinder B C,
that weighs really 41 ounce, muft prefs the top of this
Tub only with 14, which 14 ounce really counterpoifeth,
the 14 ounce of Stone in the Scale L. But how is it made
out, that the Mercury A B, preffeth up with 27 ounce?
For underftanding this, remember, that the VVater is 49
foot high, taking in the Preffure of the Air, and that a

M   water

VVater of that deepneſs is able to ſupport 41 inches of
Mercury, every inch weighing one ounce. For if **14** of
Water, be able to ſupport one of Mercury, 49 foot, or
567 inches, muſt ſupport 41. If then, the part of the
ſurface A, be able to weigh 41, it muſt have of upward
Preſſure 27 ounces, ſeing it's counterpoiſed *de facto* only
with 14. Take notice, that in the *Hydroſtaticks*, the
word *preſſing*, or *weighing*, as really and truly ſignifies a
weighing up, as a weighing down, ſeing it is no leſs eſſen-
tial to Fluid Bodies to move upward, than downward, and
that with equal force, and weight. According to this rea-
ſoning, the Ballance ſupports 28 ounces of the Water E F,
(Imagine the ſecond Tub to be ſuſpended as the firſt) ſeing
the Cylinder of Mercury D E, preſſeth up the Tub only
with the weight of 12 ounce, which 28 ounce, really
counterpoiſeth the 28 ounce of Stone in the Scale L. But
why doth the Mercury A B preſs up with 27 ounce, and
the Mercury D E with 12? For anſwer, remember, (according to the ſixth conſideration) the ſhorter a Cylinder
of Mercury is, the ſurface upon which it reſts, is the ſtronger, and more able to preſs it up; and contrariwiſe, the
longer it is, the ſurface is the more unable and weak :
therefore A B being ſhorter, and lighter than D E, the
ſurface of Water muſt preſs it up with greater force: ſo that
if the ſaid ſurface A M, be able to preſs up the Mercury
A B with 27 ounce, it muſt preſs up the Mercury D E
only with 12 ounce. According to this rule, if the Mercury A B were 15 inches high, it would preſs up only with
26 ounce, if it were 16, with 25 : if 17, with 24 : if
18, with 23, and ſo forward. This leads us to a clear diſ-
covery of all the ſecrets here : for if the Mercury A B,
thruſt up the Pipe, with the weight of 27 ounce, then
muſt

must the Scale K, be eafed of fo much weight, and fo much
muft be fubtracted from L. Now let us imagine the Pipe
A B, to be empty both of Air, Water, and Mercury:
in this cafe 41 ounce muft be in the Scale L, to counterpoife it, feing the whole Cylinder B C, that weighs fo
much, does now really counterpoife it. Let us imagine
next, thefe 14 inches of Mercury to rife, and fill the Tub
A B : in this cafe, there happens a great alteration; becaufe
the rifing of them, are really equivalent to the fubtract-
ing of 27 ounce from the Scale L; and the reafon is, becaufe by fo rifing and filling the Tub, they thruft up the
faid Tub, and by this means eafeth the Scale K, of fo much
weight. Now this Scale being eafed, you muft of neceffity take out from L 27 ounce for making a new counterpoife.

And laftly, the Scale K muft fupport the whole weight
of the Water H I, which is 39 ounce, nothing remaining
to counterballance this downward Preffure, and confequently to eafe the Ballance. How then is it counterpoifed?
For clearing this, you muft remember that this Water,
that's really 15 foot deep, muft be reckoned (as I faid)
49, becaufe of the Preffure of the Air upon the top, that's
equivalent to 34. If then it be fo, it cannot raife Mercury higher in a Tub than 42 inches; the one being 14
times heavier than the other : fo that if 14 inches of Water, cannot raife Mercury higher than one inch, 49 foot
cannot raife it higher, than 42 inches : for as 14 inches, are
to one inch; fo is 49 foot to three foot and an half, which
is 42 inches. Now I fay, the whole weight of the Water H I, refts upon the top of the Tub, and fo preffeth
down the Scale K, to which you muft imagine this Tub,
knit by a ftring, as the former was, nothing remaining to

counterpoiſe this downward Preſſure: for the top of the
Mercurial Cylinder being raiſed as high within the Pipe, as
the ſurface of Water D G S, is able to raiſe it, the ſaid
top can impreſs no force upon the Tub within, to thruſt it
up, and ſo to eaſe the Scale K. For example, when a man
erects upon his hand a Cylinder of Timber, or any ſuch
like thing, which is the outmoſt he can ſupport, he will
not be able to impreſs any impulſe, upon the ſeiling of a
room above his head; but if ſo be, in ſtead of that taken
away, there be one lighter erected, which he is able to
command, he can eaſily thruſt up the ſeiling at his pleaſure.   Juſt ſo it is here; for the 42 inches of Mercury, being the outmoſt, that the ſurface of Water D G S is able
to bear, it cannot impreſs any impulſe therewith upon the
top of the Tub within: but eaſily can the Cylinder D E
impreſs an impulſe, and more eaſily the Cylinder A B,
ſeing they are lighter, and ſo more powerful. To evidence this a little more, let us imagine two things, firſt,
the Tub G H to be empty, as if *vacuity* were in it. In
this caſe the top of the Tub ought to bear the whole burden of the Water, and conſequently the Ballance to bear
it alſo: becauſe there is not a *potentia* within the Tub, to
counterpoiſe this *pondus*. Next, let us imagine the Tub
to be only full of Water: according to this ſuppoſition,
the Ballance cannot be in the leaſt part burdened; becauſe
the Water within the Pipe, preſſeth it up with as much
force, as the Water I H preſſeth it down: and if any
thing ſhould burden the Ballance, it would be only the
weight of the Pipe, that's not conſiderable.
    From what is demonſtrated, we ſee firſt, that though
this Experiment would ſeem to prove at the firſt, that a
heavy Body weighs as much in the Water, as it doth in the
Air,

Air, because the whole weight of the Mercury A B is
found in the scale L, yet 'tis not so, because the 14 ounce of
Stone L, doth not counterpoise any of the Mercury A B, but
14 ounce of the Pillar of Water B C. Secondly, there's here a
clear ground, for asserting a *pondus* and a *potentia* in Fluids ;
because this Tub A B, is prest down with the VVater B C,
and prest up with the Mercury within it. Thirdly, there's
here a clear ground for asserting the Pressure of VVater,
even in its own place ; because the Water B C, counterpoises by it's weight, the 14 ounce of Stone L. Fourthly, we see an excellent way for finding the weight of any
Cylinder of Water ; for whatever be the weight of the
Mercury in the Tub, the Cylinder of Water, that rests upon the top, will be of the same weight exactly ; this is
evident in comparing the weight of the Mercury G H,
with the weight of the Water H I. Fifthly, that whatever be the height, and weight of a Pillar of Water, yet
the Ballance can sustain no more of it, than the just weight
of the Mercury : this is also evident, because the scale of
the Ballance, supports no more of the weight of the Water B C, than the just weight of the Mercury A B. We
see sixthly, the further down a Pipe with Mercury goes
through Water, the greater is the Pressure it makes upon
the top of the Tub within : for put the case, this were
100 foot deep, the Mercury G H, that wants all upward
Pressure now, would press up the Tub with 40 ounce : the
Mercury D E with 55, and the Mercury A B with 70.
We see seventhly, the shorter a Cylinder of Mercury be,
it is the stronger in pressing ; and longer it be, it is the
weaker ; for there's more strength in A B, than in D E.
We see eighthly, that the strength decayes, and grows, according to *Arithmetical progression*, as 1,2,3,4 ; because

if you make the Cylinder G H 41, that's now 42, it preſſeth up with one ounce. Make it 40 inches, it will preſs up with two ounces of weight. Make it 39, it preſſeth up with three. And contrariwiſe, make the Cylinder D E 29 inches; that's now but 28, it will preſs up with 11 ounce only. (VVith 28 it preſſeth up with 12.) Make it 30 inches high, it will preſs up with 10. If it be 31 inches, it preſſeth up with nine, and ſo forward. Laſtly, make the Cylinder A B 15 inches, that's now but 14, it preſſeth up with 26 (with 14, it preſſeth up with 27.) make it 16, it preſſeth up with 25; make it 17, it preſſeth up with 24. We ſee ninthly, that in Fluids, we may make a diſtinction between a *ſuſtentation*, and an *equipondium*. 'Tis evident here, becauſe there's a perfect *equipondium* between the 42 inches of Mercury G H, and the outward Water that's 49 foot deep. But 'tis not ſo, between the ſaid Water, and the Mercury D E; becauſe the ſaid Water is able to raiſe the ſaid Mercury 14 inches higher: therefore the Water only *ſuſtains* the Mercury D E, but *counterballances* the Mercury G H. We ſee tenthly, that the *pondus* of the pillar of Water B C is counterpoiſed by two diſtinct *powers* really. The one is the 14 ounce of Stone in the ſcale L, the other is the 14 inches of Mercury A B, that as really thruſts up the Water, as the ſcale K pulls it up, by vertue of the oppoſite weight. Eleventhly, take away the Stone L, and you will find the Pipe with the Mercury A B ſink down: this happens, not becauſe the ſurface of Water on which it reſts is not able to ſuſtain it, but becauſe the 14 ounce of the Water B C, that was ſupported by the Stone, doth now preſs it down. Twelfthly, the more a Body is unequally preſſed by a Fluid, the more of the weight of that Fluid is ſenſible; and the

more

more equally a Body is preſſed, the leſs ſenſible is the
weight of that Fluid: this is evident, becauſe there's a
greater weight of the VVater H I found in the Ballance
(it takes 42 ounce to counterpoiſe it) than of the VVater
E F, which is counterpoiſed with 28 ounce: and the reaſon is, becauſe the top of the Tub H, ſupports the whole
39 ounce of VVater H I, the Mercury within the Tub,
not being able in the leaſt to counterpoiſe it, or thruſt it
up. But becauſe the Tub D E, is more equally preſſed
(the VVater E F preſſeth down with 40, and the Mercury D E preſſeth up with 12) therefore leſs weight of the
VVater E F burdens the Ballance, only 28 ounce. Hence
it is, that becauſe the Tub A B, is more equally preſſed,
than either D E or G H, there's leſs of the weight of the
VVater B C, found in the Ballance, only 14 ounce. Thirteenthly, if in the inſtant of time, while the Tubs are
thus ſuſpended in the VVater, the Preſſure of the Air
above were taken away, and *annihilated*; then firſt, the 42
inches of Mercury G H would fall down, to about 13
inches. Secondly, the 28 inches of Mercury D E, would
fall down to as many. And laſtly, the 14 A B, would
ſink down to the ſame height. The reaſon is, becauſe the
Preſſure of the Air being equivalent to 34 foot of VVater, no more would remain but 15 foot, which is the real
height, according to Z M. But 15 foot of Water, cannot ſuſtain moe inches of Mercury than about 13. And
conſequently, firſt, 14 ounce of Stone in the Ballance,
would counterpoiſe the whole Water B C. The reaſon
is, becauſe the Water B C is but of 14 ounce; and the
Mercury A B, being but 13 inches high, could impreſs no
impulſe upon the top of the Tub within, that's 14 inches
high. Secondly, 13 ounce of Stone in the Scale L, would

coun-

counterpoiſe the whole Water E F, ſeing E F is but 13
ounce. Thirdly, the ſame weight (one ounce being deduced) would counterpoiſe the Water H I, becauſe in this
caſe, it weighs but 12 ounce,

To proceed a little further, imagine the Pipe G H to
be ſuſpended by the ballance, as the Pipe A B is ; and
then a little hole opened in the top H, to ſuffer the Water to come in, till the Mercury ſubſide 14 inches, namely
from Q to O (imagine this Tub to be the other) and
then ſtop it. The reaſon why the VVater ruſheth in, and
preſſeth down the Mercury, is the force and Preſſure of it:
for the ſaid VVater, finding the Cylinder in *æquilibrio* with
the outward VVater, preſently by its own weight, caſts
the ſcales, which is eaſily done, ſeeing the ſurface G S M
ſupports as much burden as it can. But that which is more
conſiderable is this ; after the ſubſiding of the Mercury
from Q to O ; the *æquilibrium* that was between the ſcale
of the ballance, and the VVater Q R is deſtroyed : for
whereas 42 ounces were required before ; 29 will now do
it. For underſtanding the reaſon of this, conſider that
between Q and O, are 14 inches of VVater ruſhed in,
which are equivalent to one inch of Mercury. Next, according to former reaſonings, the ballance muſt ſupport 29
ounces of the VVater Q R ; becauſe in this caſe, the top
of the Pipe within, is preſſed up with the weight of 13
ounces ; which in effect, diminiſheth as much of the downward Preſſure of the VVater R Q, which before had the
burden of 39 ounces. But why is the Tub preſt up with
13 ounces ? I anſwer, becauſe the Mercury, that before was
42 inches, is now but 28, or having the 14 inches of Water
Q O above it, it is 29, therefore being ſhorter, the ſurface
G S M is the more able to Preſs it up, even with as much
more force, as it is in inches ſhorter.   In

In the second place, let in as much Water more, as will deprefs the Mercury other 14 inches, namely from O to P. In this cafe, 16 ounce of ftone will make an *equipondium*; becaufe, the 14 inches of Mercury P S, and the 28 inches of Water P O Q, being a far lighter burden by 26, than the 42 inches of Mercury, the furface G S M muft be far abler to prefs them up now, than before: and therefore, muft diminifh as much of the downward Preffure of the VVater Q R, that burdens the Ballance, as themfelves wants of weight: feing then, the wholeCylinder of Mercury, and Water together, are but equivalent for weight to 16 inches of Mercury, the top of the Tub within, muft be preft up with 26 ounce; and therefore they by their upward Preffure, muft diminifh 26 ounce of the weight of the Water R Q, that weighs 39. Laftly, let in fo much VVater, as will deprefs the laft 14 inches P S; and you will find no more weight required in the Ballance to make an *equipondium*, than counterpoifeth the fimple weight of the Tub, which is not confiderable. The reafon is, becaufe, the part S, of the furface G S M, being liberated of the burden of Mercury, and fuftaining only the VVater within the Tub, in ftead of it, this furface preffeth up the VVater within the Tub, and confequently the top of it, with as great force, and weight, as the top of the Tub without is depreffed, with the outward VVater R Q: therefore, 39 ounce depreffing the Tub, and 39 ounce preffing it up, the Ballance muft be freed of the whole weight of VVater R Q. If it be objected, that the 42 inches of VVater Q S, are equivalent in weight to three inches of Mercury; therefore the part of the furface S, being burdened with this, cannot prefs up, with as great force, as the VVater R Q preffeth down. For anfwer, confider, that the part S, is able to

support

ſupport 42 ounce of VVater, and next, that the VVater
R Q weighs but 39. Then I ſay, ſeing the 42 inches of
VVater within the Tub, weighs only three ounce, the
part S, that's burdened therewith, being able to ſupport 42,
it muſt preſs up with the weight of 39, and ſo counterballance the VVater R Q.

If it be in uired, whether or not, would the 1 ; inches of
Mercury A B fall down, a ſmall hole being made in the
top of the Tub at B ? I anſwer, they would. If it be obje-
ſted, that theſe 14 inches of Mercury, are not in *æquilibrio*,
with the Preſſure of the ambient Water, as the Mercury
GH, and therefore they cannot be ſo eaſily depreſſed by the
Water, that comes in at the ſaid hole, I anſwer, they muſt
all fall down, and as eaſily, as the other, and that becauſe of
inequality of weight between the *Potentia* of the ſurface of
VVater, and the *Pondus*. It's certain, the part A of the ſurface, cannot ſupport more weight of any kind, than 42
ounce ; but when a hole is opened in B , and the VVater
comes in, 'tis then burdened with the weight of 14 ounce
of Mercury, and with the weight of 41 ounce of V Vater ;
ſo much the VVater B C weighs, which is 55 ounce : but
a ſurface that hath only the *Potentia* of 42, can never ſupport a *Pondus* of 55, no not of 43.

It may be objected thus : Put the caſe a Cylinder of
Gold, or Braſs were ſuſpended in this VVater, as the
Pipe and Mercury G H are ſuſpended by the Ballance,
would not the Ballance ſupport the whole weight of it,
without ſupporting any part of the weight of the VVater
I H, that reſts upon the top of it, I anſwer, there's a
great difference between the two ; becauſe a Cylinder of
Gold or Braſs, ſuffers both the upward and downward
Preſſure of the VVater ; but the Mercury G H, ſuffers
only

only the upward Preſſure, being freed of the downward,
by the top of the Tub. From this Experiment of letting
in the VVater upon the top of the Mercury, we ſee firſt,
that when two Fluids are in *equilibrio* one with another,
a very ſmall weight will caſt and turn the Scales, becauſe, if
the ſixth part of an inch of VVater come in at Q, it preſently alters the hight of the Mercury from 42 inches to
leſs. Secondly, 'tis impoſſible for a ſurface of Water, to ſupport more weight, than its own proper burden; becauſe the
part S, cannot ſupport more, no not a grain, than 42 ounce.
VVe ſee thirdly, that it is as impoſſible for a ſurface of
VVater, to ſupport leſs, than its own burden; becauſe
whatever loſs of weight the Pillar of Mercury S Q ſuffers,
by the ingreſs of the VVater Q O, its made up again by
the ſame VVater. If it be objected, that the 14 inches of
VVater Q O, are not ſo heavy by far, as the 14 inches of
Mercury, that fell down. I anſwer, its true, yet the part
S, is as much burdened as before, becauſe what is wanting
in weight, its made up, and compenſed by Preſſure. VVe
ſee fourthly, that the *Preſſure* of a Fluid is a thing really
diſtinct from the *natural weight*, according to the 22 Theorem : becauſe though the 14 inches of Water Q O, are
not ſo heavy naturally as the 14 inches of Mercury that fell
down, yet the Preſſure of them upon the ſurface S, is as
much. We ſee fifthly, that 14 inches of Water, that's a
body fourteen times lighter than Mercury, may have as
much weight with them, as 14 ounce of Mercury. We ſee
ſixthly, that a Cylinder of Mercury cannot be ſuſpended in
Air, or in Water unleſs it be guarded with a Tub, to pre-
ſerve it from the downward Preſſure of that Air or Water:
for by opening an hole in Q, the Mercury ſubſides. We
ſee ſeventhly, that 'tis impoſſible for two Fluids to ſuſpend

one

one another mutually, unlefs there be a fort of *equipondi-um* between them; becaufe no fooner you deftroy the *equipondium*, between the 42 inches of Mercury QS, and the part of the furface S, by the ingrefs of the Water Q O', but affoon there arifeth a new one. We fee eighthly (as we noted before) the nearer a Body comes to be equally preffed with a Fluid, the lefs is the Prefsure of that Fluid *fenfible*: becaufe lefs weight is required in the Ballance; to counterpoife the Prefsure, and weight of the Water R Q', after the ingrefs of the Water Q O P, than after the ingrefs of the Water QO. We fee ninthly, that when a Body is equally, and uniformly preffed with a Fluid, the Preffure is *infenfible*; becaufe, after the Water hath thruft down all the Mercury from Q to S, there's no more weight at all of the Water R Q found in the Ballance. We fee tenthly, that not only in Water, the Preffure of Water may be found, but out of it, namely in the Air; as is clear from the Ballance, that fupports the Preffure of the Water R Q. We fee eleventhly, a ground to diftinguifh between the *natural Ballance*, and the *artificial Ballance*. The *artificial Ballance*, is the Ballance K L: the *natural*, is the Pipe QS. We fee twelfthly, that they keep a correfpondence between themfelves, or fome *Analogy*: for by what proportion the Water thrufts down the Mercury, by that fame proportion the *pondus* L, of the Ballance is leffened: and by what proportion the Mercury rifes in the Pipe, by that fame, is the weight L augmented in the Scale. We may fubjoyn laftly, that the eafieft way of explicating the *Phenomena* of Nature, is not always the beft, and trueft. For fome may think, it were far eafier to fay, that the Ballance fupports the Mercury AB, or D E, and not any part of the Water B C, or E F. But fuch a way would be falfe, and abfurd, and contrary to all the former Doctrine.   E X-

EXPERIMENT XII.
Figure 16.

THis Schematism represents a Water 100 foot deep, whose first and visible surface is I H K. And L M is the ground of it. C D is a piece of *brass* 30 inches high, and 12 inches in diameter, suspended upon the imaginary surface of Water A N B, which is distant from the top I H K, 25 foot. This *Brass* cannot go farder down, when demitted from H ; because it's keeped up, by the Force and Pressure of the surface of Water A N B, which I prove thus. The part B sustains *de facto*, a Pillar of Water K B 1400 pound weight : therefore the part N is able to sustain as much. I suppose here, the said piece of *Brass* to weigh 1400 pound. The Water K B is 1400 pound, because its a Pillar 25 foot high, and 12 inches thick, for one cubical foot weighs 56 pound Trois. The *connexion* of the argument is evident, because it is as easie for a surface of Water, to sustain a solid Body, as to sustain a Fluid Body : therefore, if the part B, support the Fluid Pillar K B, the part N must be able to support likewise the solid Pillar C D, which is of the same weight. If it be objected, that the part N, sustains besides the *Brass* C D, a Pillar of Water E F 22 foot high, and a half, which two will weigh 2260 pound. I answer, upon supposition, that neither Water nor Air succeeded, the space E F being void of both, the *Brass* would be suspended with the force and power of the Water N. And though this cannot be made *practicable*, yet the *Theory* of it may conduce much for explicating the secrets and mysteries of the *Hydrostaticks.* But why ought the *Brass* to be suspended

pended at 25 foot from the top? I answer, because the *potentia* of the surface A N B, is equal to the *pondus* of the *Brass*. To evidence this, consider that *Brass* is a Body *naturally* heavier then Water, I shall suppose ten times, that's to say, one inch of *Brass* will counterpoise ten inches of Water. If this inequality be, then must this Pillar of *Brass* go so much farder down, than the first surface I H K, as the one is heavier *in specie*, or *naturally*, than the other: therefore it must sink 25 foot exactly; seing a piece of *Brass* 30 inches high, requires 400 inches of Water, or 25 foot to counterpoise it: for if one inch of *Brass* require ten inches of Water, then surely 30 inches must require 300. Yet it is no matter, what the thickness be, provided it be no higher than 30 inches.

To advance some farder, let us make a second supposition, namely, while the *Brass* is thus suspended upon the surface A N B, suppose the Air to come down, and fill up the imaginary space E F, then must the *Brass* be thrust down as far as the surface O P, that's 34 foot below the surface A N D, and 59 from the top. The reason of it is this, because the weight of the Air superadded, is equivalent to the Pressure of a Pillar of Mercury 29 inches high, and 12 inches thick: therefore the *Brass* being burdened with this, it must go so farder down, till it meet with a surface, whose *potentia* is equal in weight, to the *pondus* of both, which is precisely 59 foot from the top: for if one inch of Mercury require 14 of Water, then 29 inches must require 405 inches, or 34 foot. In a word, it must go as far down, as that surface, that sustains a Pillar of Water, that would counterpoise in a Ballance, the *Brass* C D, and a Pillar of Mercury 29 inches high, and 12 inches thick, both which weighs 3290 pound.

From

From what is said, we see first, that of two heavy bo-dies differing in weight, the lighter may go further down than the heavier. This is clear, because a slender *Cylin-der* of *Gold*, in form of an Arrow, half an inch thick, and 28 inches long, weighing 28 pound ('tis no matter, though the just weight of it be not determined) will go down 35 foot in Water, before it meet with a surface, whose *poten-tia* is equal in weight to its own *pondus*; for if *Gold* be 15 times heavier *naturally* than Water, then the said Cylin-der must go down before it rest, 420 inches, or 35 foot. But a piece of *Gold* 12 inches long, and six inches thick, that perhaps will weigh 208 pound, will sink no further than 15 foot. And the reason is, because, if one inch of Gold require 15 of Water to counterpoise it, then 12 must only require 180, or 15 foot. Note, that both the bodies must go down *Perpendicularly*, and not as it were *Hori-zontally*, with their sides downmost: for if they go down after this manner, they cannot sink so far. The reason of this is also evident, because a heavy body goes so far down, and no further, till it hath thrust as much Water out of its place, as will counterpoise it self in a Ballance. That's to say, if an heavy body weigh 100 pound, it must go no further down, than after it hath thrust out 100 pound of Water. But so it is, that a piece of Gold, in form of an Arrow, going down *side-wise*, or with the two ends parallel to the Horizon, will thrust as much Water out of its place, as will be the weight of it self, before it can go down 15 or 16 inches from the top: because for every inch it goes down side-wise, it expells 28 inches of Water. In going down two inches, it expells 56. In going down three inches, it expells 84, and so forward, till it go down 15 inches, where it expells 420 inches:

but

but 420 inches amounts to 35 foot. Now, take a Cylinder of Water 35 foot high, and juft the thicknefs of the Cylinder of Gold, which I fuppofed to be of half an inch, and put them in a ballance, and you will find the one juft the weight of the other. Neither can the piece of Gold go fo far down as before, if it go down *fide-wife* ; becaufe for every fix inches it is drowned, it expells a bulk of Water 12 inches long, and fix inches thick ; therefore it muft be fufpended, before it go beyond 90 inches, or feven foot and an half : now, if fix inches give one foot, 90 inches will give 15 foot : but 15 of Water in hight, and fix inches thick, is the juft weight of it in a ballance, *viz.* 208 pound. We fee fecondly, the broader and larger the furface of a Fluid be, 'tis the more able and ftrong to fupport an heavy burden : therefore the part of a furface of Water fix inches fquare every way, will carry a far greater weight, than a part four inches fquare. Though a furface of Water 34 or 35 foot deep, be not able to fuftain a Cylinder of *Gold*, if it exceed 28 or 29 inches in hight, yet take a Cylinder of *Gold*, 10 foot high, and reduce it, by making it thicker, to the hight of 20 inches, a furface of Water little more than 24 foot deep will fuftain it. Or reduce a Cylinder 10 foot high, which requires a furface more than 100 foot deep, to a Cylinder fix inches high, a furface little more than feven foot deep will fupport it. We fee thirdly, the reafon why bodies that are broad and large, move flowlier through Air and VVater, than bodies that are more thin, and flender, though both be of the fame weight in a ballance. For example, 20 pound of Lead, long and flender like an Arrow, will go fooner to the ground of a deep VVater, than a piece of Lead of the fame weight, in form of a Platter or Bafon. The reafon is, becaufe as the body is broader, so it takes a broader part of a surface, which broader part is stronger and abler, than a narrower part, and so makes the greater resistance. The same is the reason, why a Bullet six inches in Diameter, moves slowlier thorow the Air, shot from a Cannon, than a Bullet one inch in Diameter. For the same reason, Ships of seven or eight hundred Tun, move far slowlier thorow the Air, and Water, than Vessels of less burden. Item, large and big Fowls, as Eagles, move slowlier, than small Birds, as Swallows. Yea, of Fowls of the same quantity, one may move quicklier than another, as is evident in long-wing'd *Hawks*, as *Falcons*, that by the sharpness of their Wings, move far more space in half an hour, than Kites, or Gose-Hawks, whose wings are rounder. We see fourthly, that there's no body how heavy soever, but it may be supported by the surface of a Fluid, either in Air or in Water. I grant, the strongest surface of Air, that can be had, is not able to support more weight, than a Cylinder of *Gold* 28 inches high: yet though it were as large, and broad, as a *Mill-stone*, if it do not exceed the said hight, the Air is able to sustain it. For the same cause, if it were possible to free a *Mill-stone* of the Air, that rests upon it, the Air below would lift it from the ground, and carry it up many fathoms, even till it came to a surface, equal in power to the weight of the Stone. Or, if a large *Mill-stone* were demitted from the top of the *Atmosphere*, towards the *Earth*, it could hardly touch the ground, being detained by the way, by a surface counterpoising it. Or if it did touch, through the swiftness of the motion, it would surely, as it were, rebound, and be carried up again. It is alwayes to be remembred, that in such trials, the Air is supposed not to follow, or to be united, after the Stone

O   *passeth*

paſſeth thorow. Now if the Air be able to do this, far more the VVater, that's a body a thouſand times heavier. We ſee fifthly the reaſon, why heavy bodies move ſo eaſily thorow Air, and Water, namely becauſe the parts that were divided, by the body that is moved, are preſently reunited, and cloſed again, by which means it is driven forward, the Preſſure upon the back, being as much as the Preſſure before. If this were not, no body whatſoever would be able to move it ſelf one foot forward. For example, if, when a man hath advanced one ſtep forward, the Air did not cloſe again upon his back, the force of the Air upon his belly and breaſt, would not only ſtop him, but violently thruſt him backward. We ſee ſixthly, the reaſon, why the ſame body deſcends with more difficulty thorow Water, than Air, becauſe a ſurface of Water is far ſtronger, than a ſurface of Air. We ſee ſeventhly, that a heavy body is never ſuſpended by a ſurface of Water, or Air, in going down, till once it hath diſplaced, as much Water or Air, as will counterpoiſe it ſelf in a ballance. This is clear from the *Braſs* C D, that goes alwayes down, till it expell its own weight of Water. For this cauſe, if a *Mill-ſtone* were demitted, or ſent down from the top of the Air, and never reſted, till it came within 40 fathom of the *Earth*, then ſo much Air, as is expelled by the deſcent, is the juſt weight of the ſtone. We ſee eighthly, the heavier a body be *naturally*, than Water, it goes the further down, and the lighter it is, it ſinks the leſs. For if C D were of Gold, it would go further down, than being of Braſs or Iron: and if C D were a ſtone, that's lighter *in ſpecie* than Braſs, it would not go ſo far down. This lets us know the reaſon, why thicker, blacker, and heavier clouds comes nearer to the *Earth*,

than

than thinner, whiter, and lighter.   VVe see ninthly, that
the Preſſure of the Air is determinable, even in its heigheſt
degree , and ſeemes to be the ſame in all places of the
world ; but the Preſſure of the Water is not ſo.   The
reaſon of the firſt part is, becauſe the Element of Air ſeems
to be of the ſame hight in all places , and therefore we
may know its outmoſt Preſſure, which is juſt equivalent to
the weight of 28 or 29 inches of Gold, or Mercury.   But
becauſe the deepneſs of the Sea is variable, therefore the
Preſſure is variable likewiſe.   Yet if the exact deepneſs,
of the deepeſt place were known , it were as eaſie to determine the greateſt Preſſure of it , as to determine the
greateſt Preſſure of the Air.   We ſee tenthly, that a very
ſmall weight added or ſubtracted in height , will change
and alter the counterpoiſe of a Fluid.   Becauſe if you lay
but one ounce upon the top of the braſs at F, it preſently
ſubſides accordingly : or take one ounce from it, and it riſes.
But though never ſo much weight be added to it , or ſubtracted from it in thickneſs, no alteration follows.   Therefore, though this piece of Braſs C D , that's now but 12
inches in thickneſs , were made 24, by which means the
weight would be tripled and more, yet the ſame ſurface A
N B would ſuſtain it : yet , add to it in altitude , but one
inch , and preſently it ſinks down proportionably.   This
evidently diſcovers the reaſon, why its as eaſie for the Air,
to ſupport a Cylinder of Mercury 3 inches thick, as to ſupport a Cylinder half an inch thick : and why it cannot ſupport more in height than 29 inches, and why it cannot ſupport leſs.   Now the reaſon, why a thicker Pillar, is as eaſily ſuſpended, as a thinner, is this, becauſe if a Pillar of Mercury be thicker, and conſequently heavier , than it takes a
broader , and conſequently a ſtronger ſurface of Air to reſt

upon: if it be but slender, and so but light, then it takes a lesser part of a surface to bear it up, and consequently a weaker; by which means the *Pondus* of the one, is alwayes proportionable to the *Potentia* of the other. Is it not as easie for a Pillar of stone, 6 foot in Diameter, to support another six foot in Diameter; as it is for a Pillar one foot in Diameter, to support a Pillar one foot in Diameter? But as a Pillar one foot in Diameter, cannot support a Pillar 6 foot in Diameter, neither can a surface of Air, one inch in Diameter, support a Pillar of Mercury 6 inches in Diameter. But why should a larger part of a surface be stronger than a narrower part? I answer, the one is stronger than the other, for that same reason, why a thicker Cylinder is heavier than a thinner: for what I call *strength* in a surface, its nothing else but *weight*, and what I call *weight* in a Cylinder, its nothing else but *strength*. The same thing hath two names; because the pillar of a Fluid presseth down, and the surface supports: therefore, in the one its called *pondus*, in the other *potentia*. As when two scales are in *equilibrio*, either this, or that may be called the *pondus*; or either this, or that, may be called the *potentia*. Now I say, if a part of a surface four inches broad, have as much weight or force in it, as a Pillar of Mercury four inches thick; then surely, a part of a surface eight inches broad, must have as much weight and force in it, as a Pillar of Mercury eight inches thick. But why ought a surface to succumb, when the Pillar grows in hight, and not to fail when it grows only in breadth? *Ans.* VVhen it grows in breadth, the *pondus* never exceeds the *potentia*; but when it becomes higher, then it becomes heavier. That's to say, when a Pillar grows broader, there's not one part of the surface that sustains it, more burdened than another;

seing

seing the part eight inches broad, is no more prest with a
Pillar eight inches thick ; than the part four inches broad,
is prest with a Pillar four inches thick : as eight ounce of
Lead in this Scale, is no more counterpoised with eight
ounce in the other Scale, than four ounce in this Scale, is
counterpoised with four in the other. But when a Cylinder grows in hight, the *pondus* exceeds the *potentia* ; one
part of a surface being more burdened than another. We
see eleventhly, that in a large surface of a Fluid, wherein
are many parts ; each part is able to sustain its own proper
burden. So a part eight inches in Diameter supports a
Pillar eight inches thick ; and a part four inches, supports
a Cylinder four inches thick ; but cannot support a Pillar
six inches thick. But this seems rather to flow from the
disproportion of *Magnitudes*, seing a circular plain 4 inches
in diameter, cannot receive a Base of a Pillar 6 inches in diameter. But this is certain from the very nature of Fluids,
that in a deep VVater, wherein may be distinguished 100,
or 1000 different surfaces, each one is able to support his
own burden, and no more.

EXPERIMENT XIII.
Figure 17, 18, 19.

FOr making this Experiment, take two *plain* Bodies of
Brass, or Marble well polished. Make them of any
quantity ; but for this present use, let each of them be four
inches broad square-wise. Upon the back part, let each
one have an handle about six inches long, of the same metal,
formed with the *plain* it self, in the founding (if they be
of Brass) as is represented in this Schematism. When
they

they are thus prepared, anoint their inner-sides with Oyl
or Water, and having thrust the one face alongst upon the
other, with all the strength you have, till all the four edges
agree, two whereof are represented by A B, and C D, you
will find them cleave so closs together, as if they were but
one Body. The effect is this, that ordinary strength will
not pull them asunder ; and that under a surface of Water,
a stronger pull is required than in the Air.

That we may deduce some *Hydrostatical* conclusions
from this Experiment, let us suppose these two *plain* Bodies to be united in the middle of the VVater I K P Q,
that's 34 foot deep, and suspended by a beam or long tree
T V existing in the Air, near the top of the VVater, by a
chord S E passing between the middle of the beam, and
the end of the handle at E. Suppose next a great weight
of Lead R, 350 pound, to be appended to the end of the
handle at H, of the under *plain* Body C D N O. This
done, I affirm, that the beam T V, neither sustains the under *plain* Body C D N O G H, nor the 350 pound
weight of Lead R, that hangs down from the handle G H.
If it be objected, that the beam supports the upper *plain*
Body A B L M F E; therefore it must bear the weight
also of the under *plain* C D N O G H, with the weight R;
seing they are both united together, and cleave so closs,
as if they were but one Body. I answer, it supports the
one unquestionably, but not the other. To explicate
this *Hydrostatical* Mystery, I must aver three things ; first,
that the inferior *plain* is supported by the upward Pressure
of the lower VVater P Q N O. Secondly, that the
burden which the beam sustains, is not the weight of the
under *plain*, but the weight of the 34 foot of Water I K
L M. Thirdly, that this weight is exactly the weight
of

of the inferior *plain*, and Lead R. But is it not more easie to say, that the beam supports both the *plains*? I answer, if I say so, I can neither affirm truth, nor speak consequentially, But may it not be said, that the inferior *plain* is supported both by the beam, and the lower water P Q N O? I answer, this is impossible; because one and the same weight, cannot be supported totally, by two distinct supporters.

For making these assertions evident, I must suppose the superior Water I K L M to be 34 foot deep, and to weigh, if it were put into a ballance, 400 pound: and which is unquestionable, that the said Water rests upon the back of the superior *plain* L M. I suppose secondly, that the lower Water P Q N O weighs as much, and thrusts up the inferior *plain* with as great weight, as the superior *plain* is prest down with, by the superior Water. This is evident from former Experiments. And lastly, I suppose each *plain* to weigh two pound, and the weight of Lead R 350. It is to be observed here, that no mistake may arise in the calculation afterwards, that though it be said, this 34 foot of Water weighs 400 pound, yet in it self it weighs but 200: but considering the Pressure of the Air upon I K, which is as much, it may be truly said to weigh 400. These things being premitted, I say the weight that the beam T V sustains, is not the weight of the interior *plain*, and the Lead R; but 352 pound of the superior VVater I K L M, and consequently, that the inferior *plain* is supported by the lower VVater P Q N O. The reason is, because the lower VVater presseth up with the weight of 48 pound. It is in it self 400 pound: but being burdened with 352, it cannot thrust up with more weight than 48. Now, it pressing up with 48, must ease

the

the beam of 48, and counterpoiſe ſo much of the ſuperior VVater, and conſequently the beam muſt ſupport
only 352 pound of it. But put the caſe (you ſay) the
weight R, were 130 pound, 160 pound, or 180 pound,
would the beam be leſs or more burdened with the ſuperi-
or Water? I anſwer, if R be 130 pound, then the beam
ſupports only 132 pound of the ſuperior Water ; for if
the inferior be only burdened with 130, the weight of R,
and with two the weight of the inferior *plain*, then muſt
it preſs up with 368, and by this means, muſt eaſe the beam
of ſo much, it ſuſtaining 132 pound only. According to
this compting, when the Lead R weighs 160 pound, the
beam ſupports only 238 pound of the ſuperior Water.
If it weigh 180 pound, it ſuſtains 218. And if the weight
R were taken away, the beam ſupports no more of the
ſuperior VVater than two pound.

   To proceed a little further ; imagine the two *Plains* to
be drawn up 17 foot nearer the firſt ſurface I K, namely as
high as Z W. This done, the union breaks up, and they
preſently fall aſunder. The reaſon is, becauſe the ſurface Z W is not able to ſupport 352 pound, but only 300,
which I prove thus. If 68 foot ſuſtain 400, then 51
foot muſt ſuſtain 300. I ſay 68, and not 34, becauſe as
was noted, the Preſſure of the Air upon the ſurface I K,
is equivalent to other 34 foot: and therefore though
the deepneſs of this VVater, between I K and L M be but
34 foot really, yet it is 68 foot virtually, and in effect.
Imagine ſecondly the ſurface I K to ſubſide 17 foot, namely to Z W. In this caſe the union is broken alſo, and the
lower *Plain* falls from the upper. The reaſon of this, is
the ſame with the former ; becauſe by what proportion
you diminiſh the hight of the ſuperior VVater, by that
                                                         ſame

same proportion you diminish the upward Pressure of the lower VVater. Therefore, if you subtract from the superior VVater 17 foot, that weighs 100 pound, you subtract likewise 100 pound from the inferior VVater, and consequently, you make it press up only with 300, but 300 is not able to counterpoise 352.

Let us suppose thirdly, the superior *Plain*, and the superior Water to be annihilated ; then I say , the Pressure and force of the under Water would thrust up the inferior *Plain* and the weight R about eight foot higher then X Y and there suspend them. The reason is , because the surface X Y, being able to sustain 400, and being burdened only with 352, must have the weight of 48. Now the upper *Plain* being taken away , and the upper Water also, and the empty space of both remaining , the said weight of 48 pound, must carry the under *Plain* as high as is said. Let us suppose fourthly, the Pressure of the Element of Air, that rests upon I K, to be taken away, then must the two *Plain* bodies be disunited, the inferior falling from the superior. The reason is , because in this case, the superior Water would have but the weight of 200 pound, and consequently the inferior, would press up only with as much : but 200 is not able to counterpoise 352.

From what is said we see first , that in all Fluids there is an upward Pressure , as well as a downward ; and that the one is alwayes of equal force to the other : because the inferior *Plain* is pressed up with as great force , as the superior *Plain* is pressed down with. We see secondly , that in Fluids, there is a *Pondus* and a *Potentia*. The *Potentia* here is the inferior Water , and the *Pondus* is the superior. Or, the 350 pound of Lead R, may be called the *Pondus*, which counterpoiseth the *Potentia* of the surface of VVater X Y.   We see thirdly, that though the Pressure of a Fluid, be not the same thing with the natural weight, yet it is equivalent to it : because the 352 pound of Lead R , is sustained by the Pressure of the interior VVater, which could not be, unless they were virtually the same.   We see fourthly, that there may be as much Pressure in one foot of Water, as there is weight in 100, or in 1000 foot, or in 1000 fathom.   For put the case, these two plain bodies were suspended, 100 fathom below the surface of the sea, and within a foot or two of the ground, as much weight would be required to pull them asunder, as is the weight of a Pillar of Water 100 fathom high, and 4 inches thick every way, which will be more then 3000 pound weight, besides the weight of the Air above, that will weigh 200 pound.   This could not be, unless there were as much Pressure in the lowest foot of this Water, that's 100 fathom deep, as there is weight in the whole Pillar above.   We see fifthly, the more the *potentia* of a surface is burdened, the more sensible is the *pondus* : because the heavier you make the Lead R, that burdens the inferior Water, the more weight of the superior Water rests upon the Beam.   We see sixthly, the more *unequally* a body is pressed, the more the Pressure is *sensible.* For understanding this, consider that the under-face of the superior *Plain*, is more and less pressed, according to the more and less weight the Lead R is of : for put the case, the inferior *Plain* were taken away, the face of the superior *Plain*, would be equally prest with the back of it.   But when the inferior *Plain* is united to it, the Pressure of the Water is kept off ; by which means the back is prest more than the face.   Now, as the inferior *Plain* becomes heavier and heavier, by making the weight R more and more weighty,

weighty, the leſs and leſs is the face of the ſuperior *Plain*
preſt up. Hence it is, that as this inequality of Preſſure
becomes greater and greater; ſo the weight of the ſuperi-
or Water, affects the Beam more and more. Or, if the
ſuperior *Plain* were a ſenſible body, as *Animals* are, it
would find the back of it more and more burdened, according as the weight R, becomes heavier and heavier. We
ſee ſeventhly, that Water weighs in Water: becauſe all
the weight the Beam ſupports, is the burden of the ſuperior VVater, and not the burden of the inferior *Plain*, or
of the weight R. It ſupports the weight alſo of the ſuperior *Plain*, but this is not conſiderable. This is only to be
underſtood, when the Preſſure is unequal; for if the upper *Plain* were as much preſt up, as it's preſt down, the
weight of the ſuperior VVater would not be found by the
Beam. We ſee eighthly, that the higher a ſurface be, it
is the weaker; and the lower it be, it is the ſtronger: becauſe when the two plain bodies are pulled up, 17 foot, they
fall aſunder. We ſee ninthly, the vanity of the common
opinion, that maintains two plain *bodies* to cleave cloſs together for fear of *vacuity*; and that neither *Humane* nor
Angelick ſtrength is able to break this union, without the
rupture and fracture of them both.

It may be enquired, upon ſuppoſition, that the inferior
plain had four holes cut thorow the middle, ſquare-wiſe,
as A B C D in the 18 Figure, what *Phenomena* would
follow? Before I anſwer, conſider that this Figure repreſents the inner face of the Braſs-plate C D N O, of the
17 Figure, which as was ſuppoſed, is four inches from ſide
to ſide, and conſequently contains 16 ſquare inches. Now,
imagine the under *plain* C D N O, while it is united to
the uppermoſt, to have four ſquare inches cutted out of it,

as

as A B C D. Thefe things being rightly conceived, and underftood, I fay, when the faid holes are cutted thorow, the beam T V, that now fuftains 350 pound, fhall by this means, only fuftain 250 pound. To make this evident, confider that the under *plain* (as was faid) contains 16 fquare inches. Next, that the top of the inferior Water upon which the *plain* refts, contains as many, and that every inch of the Water weighs 25 pound, feing the whole, as was fuppofed before, weighs 400 pound. Now, I fay, the beam muft fupport only 250 pound of the Water I K L M; becaufe, thefe holes being made, the top of the inferior Water comes through them, and preffeth up the face of the fuperior *plain* with 100 pound, and fo eafeth the beam of fo much. I affirm next, that though the inferior Water N O P Q be in it felf 400 pound, and confequently able to fupport the inferior *plain*, with the weight R, albeit they weighed fo much, yet the faid holes being cut out, it is not able to fupport more burden than 300. The reafon is, becaufe of 16 parts that did actually bear up before, there are only 12 now that fuftains. And every one of thefe twelve, being but able to fupport 25 pound, it neceffarily follows; that the greateft weight they are able to fuftain, is 300 pound. I affirm thirdly, that if a fifth hole were cut through, the under *plain* would fall from the upper; becaufe in this cafe, the inferior Water is not able to fupport 350 pound as before, feing of 16 parts, there are five wanting, and eleven remaining, cannot fupport more weight than 275 pound. Moe queftions of this kind might be propofed; as firft; what would come to pafs, if the the upper *plain* had as many holes cut through it, anfwering to the four of the nether? Secondly, what would follow, if the nether *plain* were intire, and four bored through

the

the upper? But I shall supersede, and leave these to be gathered by the judicious Reader.

From this Experiment we see first, that the broader and larger a surface of a Fluid be, it's the more able to sustain a burden, and the narrower it be, 'tis the less able. Secondly, that each part of a surface, is able to sustain so much weight, and no more, and no less.

Before I put a close to this Experiment, it will be needful to answer an objection, proposed by *Doctor More* in his *Antidote against Atheism*, against the Pressure of the Air, which in effect militats, by parity of reason, against the Pressure of the VVater likewise. He argues thus. If the Air were indowed with so much Pressure, as is commonly affirmed, then it ought to compress, squeez, or strain together, any soft body that it environs, as, *v. g. Butter*. Put the case then, there were a piece of *Butter*, four inches broad every way, and one inch thick, containing 16 square inches, upon every side; as may be represented by the Figure 19. In this case, there is a far greater Pressure, upon the two faces, than upon the four edges; and therefore, it ought to be comprest, and strained together, to the thinness of a sheet of Paper. For answer, let us suppose the piece of *Butter*, to be 30 or 40 foot below the surface of a Water, where it ought to suffer far more Pressure, than above in the Air. Next, that it lies *Horizontal*, with one face upward, and the other downward. Thirdly, that the upper face supports a Pillar of Water 200 pound weight, and consequently, that the under face is prest up with as much. And lastly, that every edge is burdened with 50. It may be represented, with the help of the fancy, in the 19 Figure, where A B is a piece of *Butter* four inches square, and one inch thick. Only take

notice,

notice, that nothing here is repreſented to the ſight, ſave one of the four edges, namely A B; the other three, and the two faces being left to the fancy : Yet, the upper face may be repreſented by F H K M, and the under by N O P Q. Theſe things being rightly underſtood, it is wondered, why the two great and heavy Pillars of Water, the one E G I L F H K M, that preſſeth downward, and the other N O P Q R S T V, that preſſeth upward, do not ſtrain together the ſides of the *Butter*; ſeing the Preſſure of the Water B C, and the Preſſure of the Water D A, are far inferior to them for ſtrength, even by as much difference, as four exceeds one. Though this objection ſeem ſomewhat, yet it is really nothing, which I make evident after this manner. Firſt, I grant that the upper face F H K M is burdened, with 200 pound, and the nether face N O P Q with as much. Secondly, that the edge B, is only burdened, with 50 pound, as is the edge A. The other two edges, ſuſtains each one, as much. Secondly, though this be, yet I affirm the two ſides to be no more burdened, than the edges: that's to ſay, the Preſſure upon the ſides, is equal to the Preſſure upon the edges, which I prove thus. The Preſſure upon the part M, is equal to the Preſſure upon the part K, but the Preſſure upon the edge B, is equal to the Preſſure upon the part M : therefore the Preſſure upon B, is equal to the Preſſure upon K. The major Propoſition is evident, becauſe the Pillar of Water L M, is of the ſame weight, with the Pillar of Water I K. The Minor is alſo evident, becauſe, the Pillar B C, is of the ſame weight, with the Pillar L M. Now, if the Preſſure upon the edge B, be equal to the Preſſure upon M and K, it muſt be likewiſe equal to the Preſſure upon H and F. If this be, then the

                                                              edge

edge of the *bottom* **B**, muſt **be no more preſt**, than the ſide **F B K M**: therefore the Water B C, can **no more** yield to the VVater E F G H I K L M, and ſuffer the *bottom* to be ſqueezed out at B, than the VVater L M, can yield to the VVater E F G H I K, and ſuffer the *bottom* to be ſqueezed out at M. If any men ſhall affiſt and ſay, that the upper ſide bears the weight of ten Pillars, which weigh 400 pound; but the edge B is only burdened with 50: therefore 50 ought to yield to 400. I anſwer, according to the 19th Theorem, equity, that a ſlender Pillar of a Fluid is not above preſt, or moves ſlowlier, unleſs there be an unequal Preſſure; therefore the thick Pillar, that puſheth the ſlender, cannot move the ſlender Pillar, that puſheth the edge: but there is here no unequal Preſſure, being the Water X Y Z V, is of the ſame height with the other Pillars that reſts upon the feet of the ſame. I grant, if the Grid Water were not ſo high, as the other is, by the one half; then ſurely the *bottom* would be ſqueezed out at B, becauſe the ſtronger Pillar be, the leſs Preſſure is in the ſlenderer under it; therefore, there muſt be leſs Preſſure, according to that ſuppoſition in the Water B C, then now is. Or put the ſide, the Pillar I K were ſhorter then G H, or I M, the ſame which would follow: namely, a ſqueezing out of the *bottom* from K. Or, let us ſuppoſe the Pillar K, to be higher than G H or I M. In ſuch a caſe, the weight of the ſaid Pillar would preſs through the *bottom*.

From what is ſaid, we ſhall only infer this concluſion, that equality of height between Pillars at a Field maketh equal Preſſure, and inequality of height makes unequal Preſſure. The three I ſet in matter, whether they be groſſer or finer, thick or ſlender, provided they be all of the ſame Animals.

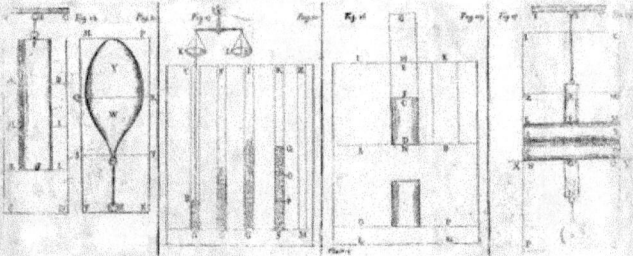

EXPE,

EXPERIMENT XIV·
Figure 20.

THis Schematiſm repreſents a Veſſel full of Water 8 foot deep. E F is a Glaſs-Pipe, open at both ends, about 9 foot high, and one inch in Diameter. A B C D is a Veſſel of Glaſs, or of any other metal, thorow whoſe orifice above, the ſaid Pipe comes down. B H I is a Pipe going out from the ſaid Veſſel, crooked with a right angle at H, that the orifice I may look upwards. That ſome *Hydroſtatical* concluſions may be inferred from this Experiment, fill the lower Veſſel A B C D with Quick-ſilver almoſt; then pour in as much Water above it, as will fill the ſpace A B H, leaving from H to I full of Air. Next, thruſt down the orifice of the Pipe E, below the ſaid Water and Mercury, till it reſt upon the bottom C D. Laſtly, ſtop well with cement the paſſage of the lower Veſſel, through which the Pipe came down, that neither Air nor Water may go out, or come in. Theſe things being done, let down this *Engine* to the bottom of the large Veſſel, which, as was noted, is full of VVater from M N to K L, 8 foot, and you will find the Mercury to riſe in the Pipe from A B to G, 6 inches, and more. The reaſon is, becauſe there is a Pillar of VVater K I, that enters the orifice I, and preſſeth down the Air, from I to P, 3 inches, which before was 6. This Air being ſo burdened; inſtantly preſſeth forward the VVater H B A: and this preſſing the ſurface of the ſtagnant Mercury A B, cauſes the liquor run up the Pipe from A B to G, 6 inches: The reaſon, why it riſeth 6 inches, is this: between the ſurface of the ſtagnant Mercury A B, and the top of the

water

Water L O K , are 84 inches. Now Water being 14 times *naturally* lighter then Mercury , there muſt be 14 inches of Water , required for ſuſtaining one inch of Mercury , and conſequently 84 , for ſupporting 6. For a ſecond trial , lift up the whole *Engine* to the top of the Water , and you will find the 6 inches of Mercury B G ſink down , and become no higher within the Pipe , than the ſurface of the ſtagnant Mercury A B without. The reaſon is , becaufe by coming up above the Water , the Preſſure of the Water K I , is taken away from the orifice I , by which means the compreſt Air H P , extending it ſelf to I , liberats the Water A B H of the Preſſure it had , and this freeth the Mercury of its Preſſure , and ſo the 6 inches falls down. For a third trial , ſtop cloſely the orifice I , and let all down as before. In this caſe , you will find no aſcent of Mercury from B to G : becauſe the Water K I cannot have acceſs to thruſt down the Air from I to P , as formerly.

For a fourth , open the ſaid orifice I , while the *Engine* is below the Water , and you will find the Mercury riſe from B to G : becauſe the Pillar of Water K I , hath now acceſs to preſs. For a fifth trial , ſtop the orifice I , and bring up all to the top , and you will find the ſix inches of Mercury B G ſuſpended , as if the *Engine* were under the Water. The reaſon is , becauſe the ſtopping of the orifice , keeps the incloſed Air P H , under the ſame degree of Preſſure it obtained from the Water K I. For a ſixth proof , open the ſame orifice I , while the *Engine* is above the Water , and you will find the ſix inches of Mercury fall down , becauſe the impriſoned Air H P , obtains now its liberty ; and expanding it ſelf from H to I , eaſes the Water B H of the burden it was under. For a ſeventh , pour in 14

Q   inches

inches of Water at the orifice F, till it reft upon the top
of the Mercury at G, and you will find one inch fall
down. Pour in as much, and two inches falls down. In
a word, pour in as much Water, as will fill the Pipe to O,
and you will find the whole fix inches fall down. The
reafon is, becaufe the Water K I, is not able to fuftain,
both the fix inches of Mercury and the Water, that's
poured in ; any one of them being able and fufficient to
counterpoife it, For an eighth trial, empty the Pipe of
the faid Water, and after the Mercury is afcended from
A B to G, as formerly, fuck out the whole Air between
G and F, and you will find the Mercury to rife from G to
R 29 inches. The reafon of this is evident from the
Pillar of Air S K, that refts upon the top of the Pillar of
Water K I : for by fucking out the faid Air, you take
away the *pondus* or weight, that counterpoifed the weight
of the Pillar S K, therefore it finding its counterpoife
removed, prefently caufeth the Water K I, to enter
farder within the crooked Pipe, till it hath preft up the
liquor to R. For a ninth trial, take the fix inches of
Mercury B G, and put them into the fcale of a ballance ;
then take as much Water, as will fill the Tub between
A B and O, and put it into the other fcale, and you
will find a moft exact counterballance between them.
The reafon is, becaufe if the Water K H, or a Pillar of
that hight, be able to raife and counterpoife the Mercury
B G; then muft as much Water, as fills the Pipe
betwen B and O, be the juft weight of it. The reafon of
this confequence is, becaufe thefe two Waters are of the
fame weight : therefore, if the one be the juft weight of
it, the other muft be fo too. If it be faid, that the
Water, that fills the Pipe between B and O, is far
                                              thicker,

thicker, then the Water K H; therefore they cannot be
both of one weight. I answer, equality of *altitude*, in this
Ballance of Nature, is equality of *weight* : therefore,
seing the one Water, is as high as the other, they must
be both of one weight. If it be said, that a Pillar of
Water between K and H, cannot counterpoise the six
inches of Mercury B G, both being put into a ballance :
and the reason is, because the one is thicker than the other.
I answer, this only proves that two Pillars differing in
weight in the *Libra* or *Artificial Ballance*, may be of one
weight in the *Natural Ballance* : because in the *Artificial
Ballance*, bodies counterpoise one another, according to
all their dimensions; but in the *Natural Ballance*, such as
this *Engine* is, Fluids counterpoise one another, according to their *altitude* only.

From the first trial, we conclude first, that Water
even in its own place *gravitats* and *weighs*, because this
Water by its Pressure, *de facto* thrusts up 6 inches of
Mercury. We see in the next place, that the Pressure of
a Fluid, is as easily communicated *Horizontally*, as *Perpendicularly*; because the Pressure runs alongst from H to B.
We see thirdly, that Fluids, may have as much Pressure
begotten in them, even while they are environed about
closely with solid bodies, whereby the superior Pressure,
immediatly and directly by perpendicular lines is keeped
off, as if they were immediatly under the Pressure: because
the Mercury A B C D, is as much burdened with the
Pressure, that comes from H, as if the upper part of the
Vessel A B, were open to let in the superior Pressure, by
perpendicular lines. The Air then under the roof of a house,
is under as great a Bensil and Pressure, as the Air without,
that's directly under the Pressure of the Atmosphere.

VVe

VVe ſee fourthly, that the Preſſure of a Fluid, may be as eaſily communicated thorow the parts of *Heterogeneous* Fluids, as thorow the parts of *Homogeneous*; becauſe the Preſſure of the VVater K I, is as eaſily communicated thorow the Air P H, thorow the Water H B, and thorow the ſtagnant Mercury B D to the orifice E, as if nothing interveened but VVater. VVe ſee filthly, that Mercury can ſuffer a Preſſure, as well as VVater or Air, becauſe the ſix inches cannot riſe from B to G, unleſs the ſtagnant Mercury A B C D were compreſſed, even in all the parts of it.

From the ſecond trial, we ſee, that there cannot be a *Pondus* in a Fluid, unleſs there be a *Potentia*, to counterpoiſe it: for when you take away the Water R I, by lifting up the Engine to the top of the Water, the Mercury B G preſently falls down. From the third trial, we conclude, that the Preſſure of a Fluid, cannot be communicated thorow ſolid Bodies: for when the *Engine* is drowned below the Water, with the orifice I, ſtopped, no aſcent of Mercury follows. We conclude from the fourth trial, that it is impoſſible for two Fluids to counterpoiſe one another, unleſs they be *in Equilibrio*; becauſe the Water K I cannot ſuſtain the Mercury B G, unleſs it be of the ſame weight. From the fifth, we conclude, that a Fluid may be keeped under the ſame degree of compreſſion, after the ſuperior weight that begat it, is taken away: for after the *Engine* is brought above the Water, with the orifice I ſtopped, the Mercury B G is ſtill ſuſpended, even by vertue of the Preſſure, that's in the ſtagnant Mercury. This tells us, that a ſphere of glaſs full of Air, may retain its *Benſil*, even though the whole Element of Air, that begat it, were deſtroyed. From the

ſixth

sixth we gather, that a Fluid cannot abide under Preſſure, when the burden is taken away that begat it, or that keepꝛ ed it under Preſſure: for by opening the orifice I, the Air P H extends it ſelf: and ſo are the VVater, and Mercury within the Veſſel freed of their Preſſure likewiſe. We gather from the ſeventh trial, that in the *Ballance* of *Nature*, one Scale cannot be more burdened then another; or that two *Fluids* cannot counterpoiſe one another, unleſs they be *in equilibrio :* for when you pour in 14 inches of Water, upon the top of the Mercury at G, they thruſt down one inch, that there may be a juſt *equipondium*, between them, and the oppoſite weight K I. We gather from the eighth trial, which was obſerved before; firſt, that there cannot be a *Potentia* in a Fluid, unleſs there be a *Pondus* to counterpoiſe it : for when you ſuck out the Air G O, which was the *Pondus*, that counterpoiſed the Air S K, this preſently in ſtead of it, raiſeth 29 inches of Mercury from G to R. We ſee ſecondly, that one pillar of Air can counterpoiſe another, Fluids of diverſe kinds interveening: becauſe the Air S K, counterpoiſes the Air within the Pipe G O, the VVater K P firſt interveening; the Air P H next interveening, and the ſtagnant, and ſuſpended Mercury interveening alſo. We ſee thirdly from this eighth trial, that the Preſſure of the *Atmoſphere*, may be communicated thorow diverſe kinds of Fluids, without the leaſt diminution of its weight: becauſe the weight of the Pillar of Air S K, is communicated, and ſent down thorow the Water K I, thorow the Air P H, thorow the VVater H B, thorow the ſtagnant Mercury B D, and up thorow the ſuſpended Mercury B G, till it ſuſpend the 29 inches between G and R, which is the juſt counterballance of it. We ſee moreover, that Fluids counterpoiſe

poiſe one another, according to altitude only, and not
according to thickneſs and breadth ; by comparing the Water K I, that's but half an inch thick, to the Mercury B G,
that's a whole inch thick. We ſee from the laſt trial, that
when a Fluid is neceſſitated, to counterpoiſe a Fluid of
another kind, in ſtead of a Fluid of its own kind, it ſuſtains
no more of it, than what is the juſt weight of the Fluid of
its own kind, becauſe the VVater K I, being under a
neceſſity to counterpoiſe the Mercury B G, in ſtead of ſo
much VVater as would fill the Tub, it ſuſtains no more of
it, than the juſt weight of ſo much VVater, as is ſaid.
We ſee ſecondly, that when two Fluids of divers kinds,
do counterpoiſe one another, that which is heavieſt in
ſpeciè, hath alwayes the ſhorteſt Cylinder. Next, that
the difference between their altitudes, is moſt exactly according to the difference between their natural weights,
therefore B G is 14 times lower than B O ; becauſe Mercury is 14 times heavier than VVater. We ſee moreover,
that though two Cylinders of a Fluid, can counterpoiſe
one another in the Natural Ballance, ſuch as this Engine
is, yet they will not do it in the Artificial Ballance : becauſe though B G counterpoiſe K I in this Ballance, yet in
a pair of Scales, the Mercury will be as heavy again as the
VVater. We ſee laſtly, that notwithſtanding of this,
yet ſuch a thing may be ; for if the orifice I, were made as
wide as the orifice F, that the Cylinder KI might be equal
to the Mercury B G in thickneſs, then ſurely the one
would counterpoiſe the other in the Libra or Artificial Ballance.

EXPE-

EXPERIMENT XV.
Figure 21.

THis Schematism represents a Water 72 foot deep, as C D A B, together with a crooked Pipe of glass I N H, the one half whereof is I P, 56 inches high, and one inch wide, the other half is P N R H, of a far narrower diameter, with an orifice H. There is also an orifice at L, with a neck, about which is knit a small chord M L, for letting down this *Engine* to the bottom of the VVater A B. For trials cause, fill the wide glass with Mercury from P to K, and you will find it rise in the narrow Pipe, as high as the orifice H. This being done, close hermetically, or with good cement the orifice L ; then by help of this chord, let all go down from the surface C D, till it be exactly 17 foot from the top, and you will find the Mercury thrust down in the narrow Pipe, from H to R, 14 inches and an half. Let it down next, as much, and the Mercury will be yet further thrust down, namely from R to N, the part H R N being full of Water. For understanding the reason of this, consider that between N and E, are 34 foot : for so high is the slender Pillar of Water, that comes from the top, and entring the orifice H, comes down thorow the Pipe to N. Consider next, that between the said Pillar of Water, and the Mercury N P K, there is a counterpoise : but this counterpoise cannot be, unless the Pillar of Water be 34 foot high, seing between N and K are 29 inches of Mercury ; for each inch thereof requires 14 of Water. Upon this account it is, that when the glass is 17 foot drowned, 14 inches and an half are thrust down from H to R. If it be objected, that the *Pressure* and *Bensil* of the inclosed Air

I K,

I K ; is equivalent to the weight of other 29 inches ; and
therefore the Pillar of Water E H R N , muſt be 68 foot
high, before a counterpoiſe can happen.   I anſwer, 'tis true
that's ſaid, but you do not conſider, that there is a Pillar of
Air F E, reſting upon the top of the Pillar of Water, that
makes a compenſation exactly.   To ſpeak then truely and
really, the 29 inches of Mercury N P K, have the weight
of 58 inches ; and the 34 foot of Water E H R N , have
the weight of 68 foot.

For a third trial, let down the glaſs 6 foot further, and you
will find the Water pierce up thorow the thick Cylinder of
Mercury P K , and reſt upon the top K.   The only difficulty is to determine, how much will ſpring up before the
motion of it ceaſe ? 'Tis evident, that the Water will aſcend, becauſe coming to the Baſe of a thick and groſs Cylinder, that it cannot intirely lift, it muſt pierce thorow it,
ſeing the force of ſuch a Pillar of Water , is now much
ſtronger, than the Mercury : for in effect , the glaſs being
drowned 6 foot further, the Pillar that comes down thorow
the ſlender Pipe, hath the juſt weight of 34 inches of Mercury : but 29 cannot reſiſt 34 : therefore the Water not
being able to lift it , by reaſon of the diſproportion that's
between the thickneſs of the one, and the ſlenderneſs of the
other, it muſt pierce up thorow it. For clearing this difficulty , conſider , that this glaſs cannot go down from one
imaginary ſurface to another, v. g. from 34 foot, where it
was, till it come to 40, where it now ſtands, but there muſt
be an alteration in the æquipondium, ſeing by going down,
the Pillar of Water E H R N grows higher , and conſequently heavier ; and therefore, ſome VVater muſt pierce
up thorow the Mercury , for making a counterpoiſe ; for
'tis impoſſible for two Fluids to counterpoiſe one another,
                                                                        unleſs

unless they be in *equilibrio*. Consider secondly, that after
the Water is come to the top of the Mercury at K, it will
find difficulty to find a room for it self, seing the space between S and I is full of Air. Notwithstanding of this, it
must ascend. I say then, after the glass is gone down from
34, to 40 foot, there will be about four inches of VVater above K, which have reduced the 29 inches of Air
K I, to 25, S I.

If it be asked, between what two things is the *equipon-*
dium now? I answer, the first was at R, between E H R,
and R N P K. The second was at R, between N R H E,
and N P K. The third is now at S, between the 25 inches
of inclosed Air I S, as one Antagonist, and the four inches
of Water S K, with the 29 inches of Mercury K P, and
the Water P N R H E, as the other. To make a fourth
equipondium, sink the Glass other six foot, till it be 46
foot from the top C D, then must some more VVater
spring up thorow the Mercury; this of necessity must be,
seing the Cylinder of VVater N R H E, is six foot higher,
and so far heavier, than it was: if this be, then must the
25 inches of Air I S, be reduced to less quantity; seing 'tis
impossible, for one Fluid to become heavier, unless its
opposite and *antagonist* become heavier too, for an *equi-*
pondiums sake. Note, that the Air I S, will not lose
other four inches, with this six foot of VVater, as it did
with the former. The reason is, because, if for every six
foot the Glass goeth down, the Air were comprest four
inches, it were easie at last to reduce it to nothing: for if
six reduce it to four, and 12 to eight, 38 ought to reduce
it to no inches, which is impossible. Therefore I judge it
must suffer compression, by a certain proportion, as we
see upon a Scale, the divisions of *Artificial* or *Natural*

R   *Sines*

Sines grow lefs and lefs, there being more fpace between 1 and 2, than between 2 and 3; more between 2 and 3, than between 3 and 4, and fo upward till you come to 90. Therefore the fecond fix foot, muft reduce the 25 inches, not to 21, but to 23 *circiter*, and fo forth. By the which means, though the Glafs fhould go down *in infinitum*, yet the Air fhall never be reduced to nothing, and there fhall ftill fome fmall quantity of VVater come up. Or in fuch a cafe, the Air may be fo compreft, that it can be no more, all the *diffeminate vacuities* being expelled. But fuppofe this to be at 1000 fathom, then at 1500, wheie the Pref-fure is ftronger, there can be no *equipondium*, which is ab-furd, for where the *pondus* becomes ftronger, the *potentia* ought to grow ftronger likewife. I anfwer, the motion of condenfation ceafeth indeed; but there ftill remains a *po-tentia*, or rather in fuch a cafe, a perfect *refiftentia*, where-by the Air is able to refift the greateft weight imaginable, before it can be reduced to nothing, or fuffer a penetration of parts, that's to fay, two parts to be in one fpace.

From the explication of thefe Phenomena we conclude firft, that in Water there is a confiderable Preffure, feing in letting down the Glafs 17 foot, the Mercury is preft down from H to R, and from R to N, in going down other 17 foot. Secondly, that 29 inches of Mercury are as heavy as 34 foot of VVater: becaufe the Mercury K P N makes a juft *equipondium* with the VVater E H R N. Thirdly, that Fluids not only of the fame kind, but of different kinds, do counterpoife one another according to *altitude*, and not according to *thicknefs*; becaufe though the Mercury K P N be far thicker, than the VVater E H, yet they counterballance one another, becaufe a proporti-on is kept according to their *altitudes*. Fourthly, that a

Fluid

Fluid *naturally* lighter, may move a Fluid *naturally* heavier, and thruft it out of its own place, becaufe the Water coming in at H, thrufts down the Mercury to R, and from R to N, and fo forth. Fifthly, that of two Fluids unequal in ftrength, debating together, the weaker of neceffity muft yeeld to the ftronger, though the weaker be far heavier naturally than the ftronger, as is evident in the Mercury, that yeelds to the Water. Sixthly, that it is impoffible for two Fluids, fo long as they are unequal in ftrength, to ceafe from motion, till they come to an *equipondium*; becaufe the Water alwayes fprings up thorow the Mercury, till an equal Ballance happen. Seventhly, that one Fluid of this kind, can counterpoife another Fluid of the fame kind, though there be divers Fluids interveening: becaufe the Air F E, counterpoifeth the Air I K, or I S, notwithstanding of Water and Mercury interveening. Eighthly, that there may be as much Preffure in one inch of a Fluid, as in a million; becaufe the 29 inches of Air I S, have as much *Benfil* in them, as is in the whole Pillar of Air E F, that goeth up from the top of the Water, to the top of the *Atmofphere*. Ninthly, that when one Fluid is under Preffure, the next muft be under the fame degree of Preffure, though they be not of the fame kind, but of different forts; becaufe the Air I S, the Water S K, and Mercury K P, are furely under the fame degree of Preffure; otherwife the motion could not end. Tenthly, that when two Fluids of divers kinds do prefs one another, that which is *naturally* lighter, afcends alwayes to the higher place, and the heavier to the loweft: becaufe the Air I S, is above the Water S K, and the Water S K is above the Mercury. Note, that this is not univerfal, but only happens when the lighter Cylinder, is flenderer than the other, for if the

R 2   Mercury

Mercury K P, were no thicker than the Water P N R H, this would raise it intirely. Eleventhly, that the compreffion of Air to lefs fpace, is not according to *Arithmetical progreffion*, 1, 2, 3, 4, 5, but according to fome other proportion, which may be called *Uniform-difform*. Note here, that though this be true of the Air, while it is compreft from a more quantity to a lefs, as here, or in a *Wind-Gun*; yet it is not true of the Preffure of the Element of Air, which is more and more from the top of the *Atmofphere* to the *Earth*, according to *Arithmetical Progreffion*, as in Water. We fee laftly, that the heavieft of Fluids, fuch as Mercury, prefs upward, as well as downward; becaufe the top of the Mercury K, thrufts up the Water K S, as well as it thrufts down the Water P N R H. It may be enquired here, how far this Glafs would go down, before the 29 inches of Air I K were reduced to one inch? I anfwer, its hard to determine; but it feems it ought to go down more than 300 fathom. In this cafe, there would be 28 inches of Water above K. Let us fuppofe the orifice H to be ftopped at that deepnefs, and the Glafs brought above the Water; then, when the faid orifice is opened in the Air, you will find the whole VVater P N R H thruft out: and not only this, but the whole Mercury P K, fpring out at the orifice H likewife, except a little that remains between N and H: the reafon is, becaufe the 29 inches of Air, being reduced to one, would be under a very great *Benfil*; therefore the weight being taken away that begat it, of its own accord, it would expand it felf to its old dimenfions; which it could not do, unlefs both the 28 inches of VVater, that's fuppofed to be above K, and the Mercury K P were thruft out of their places.

EXPE-

EXPERIMENT XVI.
Figure 22.

This Schematism represents a vessel full of VVater 84 inches deep, namely from L N the first surface, to M R the bottom. From M to R in breadth are 20 inches. There are here also two Glass-Pipes open at both ends; the one, two inches wide, the other half an inch wide. Both of them are 85 inches long. X Y O is a surface of stagnant Mercury, among which the two ends of the Pipes are drowned. E C is a Pillar of Mercury six inches in height; and so is G D, both of them raised to that altitude, by the Pressure of the Water upon the surface X Y O. The Pillar E C A is supported by, and rests upon, the imaginary Pillar A P. And so is the Pillar G D B, supported by the Pillar B Q. There are three things that occurres here from this operation of nature to be enquired after. First, why ought the Mercury to rise in the two Tubs, after the Vessel is filled with Water? Secondly, why rather six inches, then seven or eight? Thirdly, what's the reason, why it rises as high in the wide Tub, as in the narrow? I answer, the Mercury rises from C to E, and from D to G, by the Pressure of the Water, that rests upon the surface X Y O. Before that the Water is poured into the Vessel, there is here a most equal and uniform Pressure upon the surface X Y O, both without and within the Tub, namely from the Air that rests upon it. But no sooner is the Water poured in, but as soon the Pressure becomes unequal; the parts of the surface without the Tub, being more burdened, then the parts C and D within. Therefore, the part that's

less

leſs preſt, muſt riſe and climb up, till the Preſſure become
equal : for it's impoſſible that a Fluid can ceaſe from motion, ſo long as there is inequality of weight between the
pondus and the *potentia*. If any doubt, let him pierce the
ſide of the Veſſel, and when the whole Water is run out,
he will find E C and G D to have fallen down, which
clearly proves the climbing up of the Mercury, to depend
upon the in-pouring of the Water. For underſtanding
the reaſon of the ſecond, remember that Mercury (as we
have often noted) is counted 14 times heavier then Water ; therefore E C muſt be ſix inches , ſeing X Y O is
preſt with the altitude of 84 inches of Water. It would
be judged no marvel, to ſee the Mercury riſe from C to E,
and from D to G , provided the face of the ſtagnant Mercury were as high as Z F. No more ſtrange it is , to ſee
the two Mercuries riſe , with the Preſſure of the Water ;
for in effect and really , the ſaid Water is the juſt weight
of as much Mercury as would fill between X O and Z F.
For underſtanding the third, remember (as was noted before) that Fluid Bodies counterpoiſe one another , only
according to *altitude* : therefore 'tis no matter , whether
the Tubs be wide or narrow. If it be enquired , how can
one and the ſame Water, counterpoiſe two Fluids of different weights ? To ſay, that Fluids counterpoiſe one
another according to *altitude*, doth not clear the difficulty ;
for it ſtill remains to be asked, why they counterpoiſe one
another after this manner ? Therefore it ſeems, that if
the Water raiſe the Mercury from C to E in the wide Pipe,
it muſt raiſe it in the narrow one from D to K. For
anſwer, conſider firſt, that as there are here two Pillars of
Mercury C E, and D G within the two Tubs , ſo there
are here alſo two Pillars of Mercury A P and B Q , under
the

the two orifices, upon which the said two Pillars stand, and rest. Consider secondly, that the *Potentia* or force of the Pillar A P, is just equal to the *Pondus* of the Pillar E C A: Item, that the *Potentia* of the Pillar B Q, is equal to the *Pondus* G D B. Thirdly, that the *Potentia* of A P. is most exactly equal to the *Potentia* of B Q; and the reason is, because their tops A and B, are parts of the same horizontal surface. I say then, if A P be equal to E C A, and B Q equal to G D B, and A P, and B Q, equal among themselves, then must E C A be equal to G D B. The same Water then, doth not counterpoise two Bodies of different weight. I grant E C A to be far heavier, than G D B, while they are weighed in a pair of scales, but the one is not heavier than the other, as they are weighed in this *ballance* of *nature*.

From what is said, we see first, that in VVater there is a Pressure, and a considerable weight. This is evident from the rising of the Mercury. VVe see secondly, that Fluids counterpoise one another, only according to *Altitude*. Thirdly, that when a lighter Fluid presseth up a heavier, there is no more prest up of it, than is the just weight of the pressing Fluid, because the Mercury E C, is just the weight of the VVater that presseth upon X Y O. That's to say, the part of the surface C, is no more prest with the Mercury E C, than the part X, is prest with the VVater L Z X. Fourthly, if Mercury were 28 times heavier than VVater, only three inches would be prest up: if it were but seven times heavier, the altitude would be at S, 12 inches above C. Fifthly, it's as easie for a large part of a surface, to sustain a large Pillar, as 'tis for a narrow part, to sustain a narrower Pillar: because A P sustains E C A, as easily, as B Q sustains G D B.

Sixthly,

6.   Sixthly, that in Fluids there is a *pondus* and a *potentia :* as is clear from the *potentia* of A P, that ſuſtains the *pondus* of E C A. The VVater likewiſe that ſuſtains, hath a *potentia*, and the Mercury E C is the *pondus* of it. Se-

7.   venthly, that there is alwayes equality of weight between the *pondus* and the *potentia.* So is the *potentia* of A P,

8.   equal to the *pondus* E C A. Eighthly, that the *pondus* begets the *potentia.* So the weight of the VVater, be-gets the *potentia* that's in A P. For make this VVater deeper, and you augment the *potentia* of A P. If you ſubtract from it, the *potentia* of A P grows leſs by pro-portion. Or the weight of E C A, may be ſaid to beget the *potentia* of A P. To proceed a little further, let us ſuppoſe the Air H E to be removed. In this caſe, the Mercury riſes 29 inches higher than E, or 35 above C; even as high as S. In the narrow Tub it will climb up to K, if you take away the Air I G. This comes to paſs, by vertue of the Preſſure of the *Atmoſphere*, that reſts upon L N. From this we gather ninthly, that there is

9.   a counterpoiſe between the Air H E, and the weight of the Air that reſts upon L N; and that a ſlender Pillar of Air, is able to counterpoiſe a thicker : for H E is far nar-

10.   rower than L N. Tenthly, that the Preſſure of the Air, can be communicated thorow divers kinds of Fluids; be-cauſe the weight that reſts upon L N, is ſent down tho-row the VVater L Z X, and down thorow the ſtagnant Mercury, and thruſts up the Liquor from A to S, 35 inches.

11.   Eleventhly, that a lighter Fluid may be made to preſs with greater burden, than a Fluid *naturally* heavier; becauſe the weight of the Air upon L N, raiſes 29 inches of Mercury,

12.   but the VVater raiſes only ſix. VVe ſee twelfthly, that Fluids have a ſphere of activity, to which they are able to

preſs

push up themselves, or Fluids of different kinds: because B B, the lightest Mercury can raise it self no higher within the Pipe, than is without. Next, the 8½ inches of Water, overrule the Mercury no higher than E. Lastly, the weight of the Atmosphere, can raise the Mercury no higher than S, 29 inches above E.

For another end, take out from among the Water, the two Pipes, and stopping closely the two outer orifices, fill them with Mercury to the brim. Then thrust them down as before, and open the said two orifices, while they are below the surface X Y O, and you will find the whole Cylinder fall down from H to E, and there half: and the whole Cylinder in the nether Pipe falls down from I to G. Or, if you please, before this be done, stop closely the orifice H, and the orifice I, and you will find the Mercury go no farther down than S, by opening the orifice A; and no farther down than K, by opening the orifice B. This leads us to a clear discovery of the reason, why the Mercury subsides, and sinks down from the top of the Tub in the Barometer, to the 29th inch, whatever the diameter of the Pipe be. And this lets us see, that the Mercurial Cylinder is suspended by the Air, after the same manner, that the Mercury E C A suspends itself: and that there is no more difficulty in the one, than in the other.

Figure 23, 24.

This Schematism represents a Water 30 fathom deep. Under the first surface A, there are six imaginary, as B C D E F G, every one whereof, is five fathom below another.

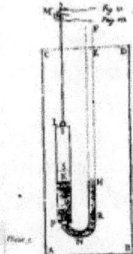

another. There are here likewife two Glaffes, each one
12 inches high, and 5 inches broad, like unto thefe,
wherein Wine,Sack, or Brandy is preferved. The Glafs
G M hath its orifice G upward. The other Glafs is compleatly open below, without a narrow orifice. For making
Experiment, take a long chord, as long as the Water is in
deepnefs, and knit the end of it round about the neck of
the Glafs at G. Take another line of the fame length,
and faften it to the bottom of the other Glafs at L. Next,
for finking the two Glaffes, take two weights of Lead,
and faften the one to the bottom at M, and the other to
the open part of the Glafs at S, and T. The two weights
then, are P and Q, each one of them about 10 or 12
pound weight. Thefe things being done, let firft down
the Glafs G M, till the weight Q fink it five fathom,
namely from A to B, and if you pull it up, you will find
the bottom covered with Water, from M to I, about four
or five inches. Let it down next, from A to C, ten fathom, and you will find more Water in it; even as much
as fills it from M to 2, about feven or eight inches. In
paffing from C D, the Water rifes from 2 to 3. If
you fink it, from D to E, the VVater rifes from 3 to
4. The VVater rifes from 4 to 5, when the glafs
is come the length of F. And laftly, when the Glafs is at
G, the loweft fathom, the VVater is as high as K. Let
down next, the other Glafs from A to B, and you will
find the Water rife in it from H to 1, four or five inches,
as in the other Glafs. In going down from B to C, it
rifes from 1 to 2. From C to D, it rifes from 2
to 3. From D to E, it rifes from 3 to 4, and fo forward, till the Glafs come to the loweft fathom, where
the Water rifes as high as I.

There

There are here several *Phenomena* to be considered. *Reflection*
First, that the Water *creeps* in at the orifice G, and fills 1.
the under part of the Glass from M to K. Secondly, that 2
not one particle of Air comes out, all the time the VVater
is in going in. Thirdly, that this Air is comprest from 3.
M to K, nine inches. Lastly, that the ingress of the Wa- 4
ter, is according to *unequal proportion* : because while the
Glass passeth from A to B, more VVater *creeps* in at G,
and fills the bottom, then in passing from B to C. And
more in going down from B to G, than in going down
from C to D, as is clear from the unequal divisions 1, 2,
3, 4, 5, 6, For understanding the reason of the first, re- *Reason*
member that in this deep Water, there is a Pressure, and
that this Pressure grows, as the VVater grows in deepness. It is then by vertue of this, that the VVater *creeps*
in, and fills the bottom of the Vessel: for in effect, every
part being under a burden, and being therefore desirous to
liberat themselves from it, they take occasion to thrust in
themselves, finding, as it were, more ease here, than
without, the Air within the Glass, being under less Pressure, than the VVater without. The second *Phenomenon* 2
is caused by the straitness and narrowness of the hole G:
for this entry being no wider, than the thickness of a Sack-
Needle, the Air cannot go out, while the VVater is coming in; that is, the passage is so strait, that the one cannot go by the other. This leads us to the reason of the
third, for if not one *particle* of Air go out, all the while
the Glass is in going down, then surely, the VVater filling between M and K, must compress the Air, and reduce
it from twelve inches to three. But the greater difficulty
is, why the ingress of the VVater is according to *unequal*
proportion. For understanding this, consider, that this

inequality

inequality, is not caused by any unequal Pressure that's in the VVater; for if this were true, then there ought to be less Pressure in the surface F, than in the surface E, and less in E, than in D, which is false and absurd. This inequality then, must flow from the nature of the Air it self, that naturally suffers compression after such a manner. 'Tis evident from the compression of Air in *Wind-guns*; for less force is required to compress the first span, than to compress the second: or contrariwise, more strength is required, to compress the third span, than the second; more to compress the fourth, than the third, and so forth. 'Tis evident in all bodies endowed with Bensil, as in the *Spring* of a *Watch*, that requires more strength to bend it, in the end, than in the beginning.

For a second trial, pull up from the bottom of the Water the Glass L I H, and when it comes above, you will find nothing in it. The reason is, because the Vessel being open between T and S, the whole VVater I H, falls down by degrees; but in effect, is really thrust out, by the strong *Bensil* of the comprest Air I L, that now expands it self, when it finds the Glass go up thorow the VVater, whose Pressure is less, and less from the bottom to the top, but the contrary effect follows, when the other Glass is pulled up; namely, the VVater remains within the Glass, and the Air above it, is thrust out by degrees, as the Glass comes nearer to the top. For understanding the reason of this, consider first, that while the orifice G, is level with the lowest surface, where it now is; that's supposed to be 30 fathom deep, there is a real counterpoise between the inclosed Air G K, and the ambient VVater without: for with what force the one strives to be in, with the same force the other endeavours to be out; and because they are in

equal

equal terms, therefore the one cannot yeeld to the other.
If you pleafe to give the victory to the VVater, then let
the Glafs go further down: but if you defire the Air to
overcome, then muft the Glafs be pulled up. Pull it then
up from the place it is in, till it come to F, and you will find
a confiderable quantity of Air come out at G, and after 2 or 3
minuts of time, emerge and come to the top A, in form of
round Bells, or Bubbles. The deepnefs and grofenefs of the
Water thorow which the Bubbles come, makes their motion
fo flow. The reafon of this eruption, muft be lefs Preffure of
Water in the furface F, than in the loweft G, from whence
the Glafs came. Suppofe then, the loweft to have fix degrees
of Preffure, F to have five, E to have four, D three, C two,
and B to have one: and fuppofing the inclofed Air K G, to be
equal in force to the Preffure of the loweft fathom, it muft
then have fix degrees of *Benfil* in it. Put the cafe then,
that with fix degrees of *Benfil*; it come to the furface F,
that hath but five, it muft furely break forth, and overcome the force and *power* of that furface: for 'tis impoffible
that two Fluids can be unequal in force and power, but
the ftrongeft muft overcome, and the weakeft yeeld:
therefore, when the orifice comes to F, the Air being
ftronger than the Water, breaks forth; and as long
doth this eruption continue, as inequality of power continues between the one and the other. In pulling up the
Glafs from F to E, other five fathom, more Air comes
out. The reafon is the fame, namely lefs Preffure in E
than in F: therefore; when the inclofed Air, that hath
five degrees of *Benfil*, comes to E, that hath but four, it
muft overcome, and fo long muft it be victorious, till by
expanding it felf, it be reduced to the *Benfil* of four. In
pulling up the Glafs from E to D, more Air yet breaks
out,

out, becauſe a ſurface of three degrees of Preſſure, is not able to reſiſt four degrees of Benſil. In paſſing from D to C , more Air comes yet out for the ſame reaſon, till in going up to the top, where there is no Preſſure , no more Air breaks out.

'Tis to be obſerved firſt, that the motion of the Air up thorow the Water is but ſlow, the medium being thick, and groſs. Secondly, that if the Glaſs be pulled up quickly, from one ſurface to another, or contrariwiſe , let down quickly, it preſently breaks in pieces. This comes to paſs through the ſtrong Benſil of the incloſed Air , that muſt have time to expand it ſelf, otherwiſe it breaks out at the neareſt : for it being of ſix degrees of Benſil, and coming quickly to a ſurface of five, there happens an unequal Preſſure , the ſides of the Glaſs being thruſt out, with greater force, than they are thruſt in with. But if ſo be, the Glaſs move ſlowly up, the incloſed Air gets time to thruſt it ſelf out by degrees, ſo that whatever ſurface the Glaſs comes to , there is little difference between the Preſſure of the Water , and the Benſil of the Air. The reaſon why the Glaſs breaks in pieces, while it goes quickly down , is likewayes unequal Preſſure upon the ſides : for in paſſing quickly from a ſurface of five degrees, to a ſurface of ſix, the ſides are preſt in with greater force , than they are preſt out with, and the reaſon is, becauſe through the ſtraitneſs of the hole G, the Water cannot win in ſoon enough, to make as much Preſſure within, as there is without. 'Tis to be obſerved thirdly ; that if the orifice G be ſtopped , before that the Glaſs be ſent down, it will not go beyond three or four fathom , when it ſhall be broken in peices; though the motion were never ſo ſlow: and this comes to paſs, through the ſtrong Preſſure

of

of the Water. Fourthly, the stronger the Glass be in
the sides, it goes the further down without breaking:
therefore a round *Glass Bottle*, will sink 20 or 30 fathom,
before that it be broken with the Pressure of the Water.
If a Vessel of iron were sent down, it ought to go much
further. An empty *Cask*, or *Hogshead*, will not sink
beyond seven or eight fathom, without breaking, or
bursting; yet a Bladder full of wind, knit about the neck
with a Pack-Threed, will go down 100 fathom, yea
1000 without bursting.

It may be here inquired, what sort of proportion is
keeped by the *unequal ingress* of the Water? I answer, it
may be known after this manner. Let first down the
Glass one fathom, and having pulled it up again, measure
the deepness of the Water in the bottom, of it. Next,
having poured out that Water, let it down two fathom,
and pulling it up, measure the deepness, which you will
find more, than afore. Do after this manner, the third
time, and the fourth time, till you come to the lowest
fathom, and you will find the true proportion.

From what is said we see first, that in Water there is a
Pressure, because through the force and power of this
Water, the 12 inches of Air that filled the Glass, are
reduced to three. Secondly, that this Pressure growes,
as the Water growes in deepness: because there is more
Pressure in B, than in A, more in C, than in B; and so
downward. Thirdly, that when Air is comprest, by
some extrinseck weight, the *Bensil* is intended, and grows
stronger by *unequal proportion*, as is clear from the unequal divisions, 1, 2, 3, 4, 5, 6. Fourthly, two Fluids
cannot cease from motion, so long as the *potentia* of the
one, is unequal to the *pondus* of the other: this is evident
from

from the Water's creeping in at G, all the while the Glaſs
is in going down ; and from the Air's coming out, all the

5 . while the Glaſs is in coming up. Fifthly, that no ſooner
two Fluids come to equality of weight, but as ſoon the
motion ends: becauſe, if the Glaſs halt at D, E or F, in the
going down, upon which follows a counterpoiſe, then

6 . doth the creeping in of the Water ceaſe. Sixthly, there
may be as much Preſſure in a ſmall quantity of a Fluid, as in
the greateſt : becauſe there is as much *Benſil* in the ſmall
portion of Air, included between K and G, as there is of
Preſſure, and weight, in this whole Water, that's 30

7 . fathom deep. Seventhly, that the *Preſſure* of a Fluid,
is a thing really diſtinct, from the *natural weight* : this is
evident from the Preſſure of the incloſed Air G K, that's
more and leſs, as the Preſſure of the Water K M, is
more and leſs, but the *natural weight* is ſtill the ſame, ſeing

8 . the ſame quantity remains. Eighthly, one part of a
Fluid, cannot be under Preſſure, but the next adjacent,
muſt be under the ſame degree of Preſſure : this is alſo
clear, becauſe what ever degree of benſil the included Air
K G is under, the Water K M is under the ſame. Therefore, when the one is under ſix, as in the loweſt fathom,
the other is under ſix likewiſe. And when the one is
under five degrees of Preſſure, as in the ſurface F, the

9 . other is under as much. Ninthly, *Benſil* and *Preſſure* are
equivalent to *weight* : becauſe the Water K M, is as
much burdened with the *Benſil* of that ſmall portion of Air
above it, as if it had a Pillar of Water 30 fathom high

10 upon it. Tenthly, that the Preſſure of Fluids, is moſt
uniform and equal, and that two Fluids of different kinds,
may preſs as uniformly, as if they were but one : this is
evident from the ſides of the Glaſs, that are not broken
in

in pieces, by the strong *Bensil* of the inclosed Air, and
heavy Pressure of the inclosed Water; and this happens
because the Pressure without, is as strong as the Pressure
within. We see lastly, that *Water does not weigh in*
Water, because when a man lets down this Glass by the
chord, to the lowest surface, he finds not the weight of
the Water K M, that's within the Glass, but only the
weight of the Lead Q. 'Tis certain, he finds not the
weight of the Water I H; because it rests not upon the
Glass within, but is sustained by 'its own surface, the
mouth of the Glass being downward, and open. When
I say *Water does not weigh in Water*; the meaning is not,
that Water wants weight or Pressure in it, but that this
weight and Pressure is not found, as the weight and Pressure
of other bodies are found, while they are weighed in
Water. For example, a piece of Lead or Gold, hung in
the Water by a string, the other end being fastened to a
Ballance in the Air, *gravitats*, and weighs down the Scale;
and the reason is, because Lead and Gold, are naturally
and *specifically* heavier than VVater; but a piece of Metal
of the same *specifick* weight with Water, or VVater it
self, cannot *gravitat* in VVater, or weigh down the Scale
of a Ballance; and the reason is, because the surface of
Water upon which they rest, bears them up with as great
weight and force, as they press down with. If it be said,
that the Water K M, rests upon the bottom of the Glass
within; and therefore, if the man above, find the weight of
the Glass, he must find the weight of the Water within it.
I answer, the consequence is bad, because the weight of the
Water within, is sustained, and counterpoised by the weight
of the Water without, whereupon the bottom of the Glass
rests. That's to say, as there is a Pillar of Water K M within

T   the

the Glaſs, that preſſeth down the bottom, ſo there is a
Pillar of Water without the Glaſs, whereupon the bottom
of the Glaſs reſts, and which bears up both, But the greater difficulty is this, the further down the Glaſs goes,
it grows the heavier, becauſe of more and more Water,
that creeps in at G. Now 'tis certain, the weight Q grows
not heavier, therefore it muſt be the Water within the
Glaſs, that makes the increaſe of the weight; and therefore Water muſt ſtill weigh in VVater. If this argument
had any ſtrength in it, it would prove the weight of the
VVater I H to gravitat and weigh likewiſe; becauſe the
further down this glaſs goes, it grows the heavier, becauſe
of more, and more Water, that creeps up from H to I,
Now 'tis certain, the weight of Lead B grows not heavier.
Behold, the difficulty is the ſame in both, and yet it were
raſhneſs to affirm the Water I H to be found by a mans
hand, when he pulls up the Glaſs with a ſtring, ſeing it is
ſuſtained by its own ſurface, and not by any part of the
Glaſs. Though this might ſuffice for an anſwer, yet becauſe the contrary is mantained by ſome, and that with a
new Experiment to prove it, I ſhall be at ſome more pains
to vindicat the truth of what I have ſaid.

This new Experiment to prove that *Water weighs in Water*, I found in a *Philoſophical Tranſaction*, of *Auguſt* 16.
Anno *1669*. *Numb.* 50, the Invention whereof is
attributed by the publiſher, to that honorable and worthy
Perſon Mr. *Boyl*, whoſe concluſions and trials, I never
much called in queſtion, but finding this oppoſite, and contrary to what I have demonſtrated, I ſhall crave liberty to
ſay, *amicus Socrates, amicus Plato, ſed magis amica veritas*;
and ſhall therefore examine it as briefly as may be. The
words of the Publiſher are as follows.

The

The *Author of this Invention is the Noble* Robert Boyl;
*who was pleased to comply with our desires, of communicating
it in English to the curious in* England, *as by inserting the
same in the* Latine *Translation of his* Hydrostatical Paradoxes, *he hath gratified the Ingenious abroad. And it will
doubtless be the more welcome, for as much as no body, we know
of, hath so much as attempted to determine*, how much Water may weigh in Water; *and possibly, if such a Problem had
been proposed, it would have been judged impracticable.*

The Method *or* Expedient, *he made use of, to perform it, as
near as he could, may easily be learned by the ensuing accompt
of a Trial or two, he made for that purpose, which among his
Notes he caused to be registred in the following words.*

A Glass-bubble *of about the bigness of a Pullets egg, was
purposely blown at the flame of a Lamp, with a somewhat long
stem turned up at the end, that it might the more conveniently
be broken off. This Bubble being well heated to rarify the Air,
and thereby drive out a good part of it, was nimbly sealed at
the end, and by the help of the Figure of the stem, was by a
convenient Weight of Lead depressed under Water, the Lead
and Glass being tyed by a string to a Scale of a good Ballance,
in whose other there was put so much weight, as sufficed to counterpoise the Bubble, as it hung freely in the midst of the Water.
Then with a long Iron* Forceps, *I carefully broke off the seal'd
end of the Bubble under Water, so as no Bubble of Air appear'd
to emerge or escape through the Water, but the Liquor by the
weight of the* Atmosphere, *sprung into the un-replenish'd part
of the Glass-Bubble, and fill'd the whole cavity about half full;
and presently, as I foretold, the Bubble subsided, and made
the Scale 'twas fastned to, preponderate so much, that there
needed 4 drachms, and 38 grains to reduce the Ballance to
an* equilibrium. *Then taking out the Bubble with the Wa-*

ter in it, we did, by the help of a flame of a Candle, warily apply-
ed, drive out the Water (which otherwise is not easily excluded
at a very narrow stem) into a Glass counterpoised before;
and we found it , as we expected ; to weigh about four
drachms and 30 grains, besides some little that remained in
the Egg, and some small matter that might have been rarified
into vapors , which added to the piece of Glass that was bro-
ken off under Water and lost there, might very well amount to
7 or 8 grains. By which it appears not only, that Water
hath some weight in Water, but that it weighs very near,
or altogether as much in Water, as the self same portion of
Liquor would weigh in the Air.

The same day we repeated the Experiment with another seal-
ed Bubble, larger then the former (being as big as a great Hens-
egg) and having broken this under Water, it grew heavier
by 7. drachms and 34 grains ; and having taken out the
Bubble, and driven out the Water into a counterpois'd Glass,
we found the transvasated Liquor to amount to the same
weight, abating 6 or 7 grains, which it might well have lost
upon such accompts, as have been newly mentioned. Thus he.

Figure 24.

THe defign then of this Experiment is to prove
that *Water weighs in Water*; but, it feems, there
is here a very great miftake, which I fhall make out after
this manner. For which caufe, let this Schematifm 24
reprefent the Experiment already defcribed. The *Glafs-*
bubble then is E P F R. The ftem is H C: the weight
that finks the Glafs is B. The furface of Water under
which it is drowned, is A D. The Ballance to which the
Glafs is knit by a ftring is N O. And laftly E F R is
the Water that came in, and filled the half of the Bubble.

Now

Now I say, it is not the weight of the Water E F R, that turnes the Scales above, and makes an alteration in the Ballance, but its only the weight of the *Lead* B, that does it. For evincing this, confider that all heavy bodies, are either lighter in *fpecie* than Water; as cork, or of the fame *fpecifick* weight with it, as fome Wood is, or laftly heavier in *fpecie* than Water, as Lead or Gold. Now 'tis certain, that bodies of the firft fort cannot *weigh in Water,* and the reafon is, becaufe they being naturally lighter, their whole weight is fupported by the Water, and therefore not one part of them, can be born up by a Ballance above. A piece of Cork that weighs 12 ounces in the Air, weighs nothing in Water, becaufe as foon as it toucheth the furface, the whole weight of it is fupported, and therefore cannot affect the Ballance above. But bodies of the third fort, as is clear from experience and reafon, does really weigh in Water: And the reafon is, becaufe they being naturally heavier than water, their whole weight cannot be fupported by it, and therefore fome part of them muft burden the Ballance, to which the body is knit. A piece of Lead, that weighs 12 ounces in the Air, will not lofe above 2 ounces, when 'its weighed in Water; or may be lefs. But here there is no difficulty. The queftion then is, in order to bodies of the fame *fpecifick* weight with Water, as fome Wood is, or as Water is, I fay of fuch alfo, that they cannot weigh in Water; and the reafon is, becaufe they being juft of the fame weight, muft have their whole weight fupported by it; even as one foot of Water, fupports the whole weight of the foot above it. It may be evidenced after this manner. Take a piece of Wood, that's lighter in *fpecie* than Water, and add weight to it by degrees, till it become of the

fame

same weight with Water. Knit it with a string to a
Ballance, ond weigh it in Water, and you will find the
whole weight supported by the Water. And the reason
is, because, being left to it self, it can go no further down,
than till the upper part of it, be level with the surface of
the Water. Now, the whole weight being thus supported, not one ounce of it can burden the Ballance. In
a word, the Ballance can never be burdened, unless the
body that's knit to it, have an inclination to go to the
ground, when left to it self, which a body of the same
weight with Water can never have. I conclude then,
if a body of the same weight with Water, cannot
weigh in Water, neither can *Water weigh in Water*, seing
Water is of the same weight with Water. And Therefore the Water E F R, that's now within the *Bubble*, cannot in anywise burden the Ballance above ; but must be
supported wholly by the Water I K G H, upon which
the bottom of the Glass rests. If it be said, that the
Glass it self is supported by the Ballance, because 'its
heavier in *specie* than Water ; therefore the Water
within that rests upon the sides of it, must be supported
likewise by it. I answer, the whole weight of the Glass
is not supported, by the Ballance, but only a part ; the
Water I K G H supporting the other part. And this
part is just as much as is the weight of Water, that's
expelled by the Glass. Now, if the said Water support so much of the Glass, because it is the just weight of
so much Water, why should it not also, support the
Water within the Glass ? Seing the Water within the
Glass, is just the weight of as much Water, as will fill
the space E F R.

I come in the next place to shew, that it is the weight
of

of the Lead B that turns the Scales, when the VVater
comes in at C, and fills the half of the ſphere, For
underſtanding this, let us ſuppoſe firſt, the weight that's
in the Scale O to weigh ſix ounces, Secondly, that the
Glaſs takes 12 ounces to ſink it compleatly under the ſurface A D. Thirdly, the weight B to be 18 ounces,
namely for this cauſe, firſt, that 12 of it may ſink the
Glaſs; next, that the other ſix may counterpoiſe the ſix
in the Scale O. Laſtly, that the VVater within the
Glaſs weighs ſix ounces. I abſtract from the weight of the
Glaſs it ſelf, which is not conſiderable, ſeing the moſt part
of it, is ſupported by the VVater, and not by the Ballance.
Now, I ſay, 'tis ſix ounces of the weight B that makes
this alteration, and turnes the Scales, For if 12 ounces
ſink the Glaſs below the VVater, when 'its full of Air,
and no Water in it, then ſurely ſix are ſufficient to ſink it,
when it is half full. And the reaſon is, becauſe there is a
leſs *Potentia* or force in ſix inches of Air, by the one half,
to counterpoiſe a weight of 12 ounces, than in 12 inches of
Air. Therefore this Air, being reduced from 12 inches
to ſix, it muſt take only ſix ounces to ſink it. If this be,
then the other ſix ounces that now wants a party to counterpoiſe them, muſt burden the Ballance, and be ſupported
by the Scale: and therefore, to make a new *equipondium*
again, you muſt make the weight O 12 ounces, by adding
ſix to it, that it may counterpoiſe 12 of B, the other ſix
being counterpoiſed by the Air E P F. Let us ſuppoſe
next, this Glaſs to be compleatly full of VVater, and
the whole Air expelled. In this caſe the Scale O, muſt
have 18 ounces in it, for making a new *equipondium*. The
reaſon is, becauſe there being no Air in the Glaſs to counterpoiſe any part of B, the whole weight of it muſt be
ſuſtained

ſuſtained by the Ballance, and therefore in the Scale O, there muſt be 18. Now, I enquire, whether theſe 18 ounces, are the *equipondium* of the VVater within the Glaſs, or of the weight of *Lead* B? 'Tis impoſſible it can counterpoiſe them both, ſeing the VVater is now 12, and B 18. It muſt then either be the counterballance of the Water, or the counterballance of the *Lead*. It cannot be the firſt, becauſe 12 cannot be in *equipondio* with 18, It muſt then be the ſecond. Or if theſe 18 ounces in the Scale O, be the counterpoiſe of the Water within the Glaſs, I enquire what ſuſtains the weight of the *Lead* B? The weight of it, cannot be ſuſtained by the Water, becauſe 'tis a body naturally heavier than Water, it muſt therefore be ſuſtained by the Ballance, I conclude then, that *Water cannot weigh in Water*. If it be objected, that this concluſion ſeems to contradict, and oppoſe the Preſſure of the Water, that's been hitherto confirmed with ſo many Experiments, I anſwer, the *Preſſure of the Water* is one thing, and *Water to weigh in Water* is another. The firſt is, when one Pillar of Water counterpoiſes another, or when a Pillar of Water counterpoiſes a Pillar of Mercury, or is counterpoiſed by a Pillar of Air, all which is in order to the *Natural Ballance*, wherein bodies weigh only according to altitude. The ſecond is, when VVater is not counterpoiſed by VVater, or by Mercury, or by Air, or by any other Fluid; but when 'its weighed by a piece of *Lead* or ſtone in an *Artificial Ballance*, for knowing how many ounces or pounds it is of, as if a man ſhould endeavour to weigh the Water E F R by help of the Ballance above, which in effect is impoſſible.

EXPE-

EXPERIMENT XVIII.
Figure 25.

MAke a *Wooden Ark* after this following manner. The Planks must be of Oak, an inch thick. The height 40 inches. The breadth 36. Closs on all sides, and above, and open below. And because the form is four-square, there must be four Standarts of Timber, in each corner one, to which the Planks must be nailed. Four likewise upon the top, crossing the other four at right angles, to which the cover must be joyned. The sides must be plained, and the edges both plained and gripped in all the parts, that the joynings may be closs. Upon the top fasten a strong Iron Ring, as at N, through which must be fastned a Rope, of so many foot or fathom. And because the use of this Engine is for *Diving* under the Water, it must therefore be all covered over with Pitch within and without, especially in the couplings. And because this Instrument cannot sink of its own accord, it must have a great weight of Lead appended to it, for that cause, whereupon the *Divers* feet must stand, while he is in going down. The precise quantity and weight of it cannot be determined ; because it depends upon the quantity of the *Ark*, which if large, requires a great weight: if of a lesser size, requires a lesser weight. But whatever the dimensions of the *Ark* may be, the weight of the *Leaden-foot-stool* can easily be found out by trial. This Invention then, is for *Diving*, a most excellent Art, for lifting up of *Guns*, *Ships*, or any other things, that are drowned below the Water. And it is in imitation of the *Diving bell*, already found out, and made use of with success. It is called a Bell, because of the form, that represents a

V   Church-

Church-bell indeed, being round, wide below, and narrower in the top: only, the matter is of Lead. It seems, it is of this mettal, firſt, becauſe Lead is weighty, and will therefore eaſily ſink: ſecondly, becauſe it's eaſily founded, and will by this means, being of one piece, be free of rifts, and leaks: thirdly, it being of Lead, will be of a conſiderable ſtrength for reſiſting the force of the VVater, that ordinarily breaks in pieces Veſſels that are weak. I cannot well divine and gueſs the reaſon, why firſt it is round, and next narrower above, than below, unleſs, becauſe its more eaſily founded after this way, than after another. This device here deſcribed is named a *Diving Ark*; firſt, becauſe it is of Timber, and next, becauſe it ſaves a man from being overwhelmed with the Waters. I preſcribe it of *Wood*, becauſe of leſs trouble, and expence in making of it. 'Tis four ſquare, becauſe it contains under this Figure, far more Air, than if it were round; even as much more, as a ſquare Veſſel 30 inches wide, contains more than a round Veſſel 30 inches wide. Now, the more Air, that's in the Veſſel, the eaſier is the reſpiration, and the longer time is the man able to abide under the VVater, which two things are of great advantage to this Art. For if by a gueſs we reckon, how much more Air is in the one, than in the other, we will find in the *Ark*, as before it is deſcribed, 30 ſquare foot of Air, but in the *Bell*, though it be 36 inches wide, as well above, as below, yet little more than 23 will be found, which is a conſiderable difference. But far leſs muſt be in it, ſeing it's narrower above, than below. Beſides this advantage, there are others very uſeful: for being of Wood, it's more tractable. Next, ſeveral Knags of Iron may be faſtened conveniently to the ſides within, to which a man faſtning his hands, may keep his body fixed and ſure in going

ing down, and coming up. Moreover, if a man were in hazard to be confounded with fear, or lose the right exercise of his senses, and so be in danger of falling out of the *Ark*; or if his feet should slide off the *foot-stool*, and his hands fail him too, a chord knit to one of those, and fastened about his wast or middle, might bring him up, though he were dead. Then, its far easier to cut out a window or two in the sides of it, not very large, but little, as K and I, whereby, they being covered with Glass, a man may see at a distance, what's upon the right hand, and what's upon the left, and what is before. This device is of excellent use, for through the want of it, the *Diver* sees no more, but what is just below him, which sometimes, when he is near the ground, will not exceed the compass of a large Milnwheel. But if so be, three holes be cut thorow, one on every hand, and one before, he may see as much bounds, and all things in it, as if he were not inclosed, and invironed with a cover. A little schelf likewise may be fixed upon the one side or the other, for holding a Compass with a Magnetical Needle, for knowing how such and such a thing lies in the ground of the Sea. In one of the corners may hing a little bottle with some excellent spirits, for refreshing the stomach, under Water. Many moe advantages I might name, this Engine being of Timber, but shall forbear; leaving the collection of them to the ingenious Reader, and proceeds to answer some objections, that may be made against it.

First, if this Engine be made of Wood, it will not sink so easily, as being made of Lead. I answer, this difficulty is soon overcome, namely by making the *Foot-stool* the heavier: therefore how light soever it be, a weight may be found to counterpoise it in the Water. If it be judged too

too

too light in Timber, it may be lined with Lead, especially
without. Secondly, if it be of VVood, there must be
couplings and joynings in it, and so rifts and leaks in it,
through which the VVater may come. I answer, there
is less difficulty here, than in the former; because the
joynts may be made so closs in all the parts, and may be so
covered over with pitch, or with some such like matter,
that it may defie either Water to come in, or Air to go
out. Thirdly, if it be made of VVood, it will be in hazard of breaking by the force of the VVater : for oft times
its found, that the strongest *Hogshead* will burst asunder by
the Pressure of it, if they go but down 7 or 8 fathom.
I answer, this objection flows from the ignorance of the
nature of Fluid bodies. If so be then, that a man knew,
that the Pressure of VVater is uniform, most equal, and
presseth upon all the parts of a body within it alike, no
such scruple would occurre. I say then, the Ark, though
no thicker in the sides, than a thin sawen dale, will go
down, in spight of all the Pressure that's in the VVater,
not only 10, but 20, or 30 fathom, without all hazard.
And the reason is, because what Pressure soever is without,
to press in the sides, the same degree of Pressure is within
to press them out. By this means, there is not one part
of the VVater, how deep soever, to which the *Ark* may
come down, but there will be found as much force in the
Air within, as will counterballance the whole weight without, as will be infallibly demonstrated afterwards. This answers a fourth objection, namely if holes be cut out in the
sides of the Ark, in stead of windows, the force of the VVater will break the Glasses in pieces, that covers them. There
is here no hazard, though the said windows were 12 inches
in Diameter : but its not needful they be so large. It's suf-
cient,

ficient, if they be 2 inches wide: for a mans eye near to a hole, 2 inches wide, will see a great way about him. There's a neceſſity the Glaſſes be joyned in with cement, that Water may not have acceſs to come in, or Air to go out. In ſuch a caſe ther's no hazard, that the Preſſure of the VVater, will break through the windows, or break the Glaſſes; becauſe the Preſſure of the Air within, being of the ſame force with the ſtrength of the VVater without, the Glaſſes are keeped intire. It may be enquired, what hazard would follow, upon ſuppoſition a ſmall hole were pierced in the head of the Ark above, when it is going down? I anſwer, ther's not ſo much hazard, as a man would think; provided the hole be not wide, but narrow. If it be wide, not only the VVater comes in, but the Air goes out, the one thruſting it ſelf by the other. If the hole be no wider, than the point of a bodkin is in thickneſs; ther's no danger at all: for by reaſon of the ſtrait paſſage; the one cannot thruſt it ſelf by the other, and therefore neither the VVater can come in, nor the Air go out. And this comes to paſs, by reaſon, that the Air within, is as ſtrong as the Water is without. Now, if they be both of the ſame ſtrength and force, why ought the Air rather to go out, then the Water to come in; or the Water rather to come in, then the Air to go out? I am confident, though the hole were as wide, as a man might thruſt in his little finger, yet no irruption of Water, or eruption of Air would follow. This demonſtrats clearly, that though a ſmall riſt, or leak ſhould happen in the *Ark*, yet no hazard or danger would follow thereupon. If it be inquired, whither the greateſt hazard is from the ingreſs of the Water, or from the egreſs of the Air? I anſwer, ther's no danger from the coming in of the Water from above; be-

cauſe

caufe as it comes in, it falls down, and fo mingles with the
reft below.　But if the Air fhould go out, the *Ark* fills prefently full of Water, and drowns the man that is in it.

The next thing confiderable in this *Diving Inftrument*,
is the foot-ftool of Lead C D, that's not only ufeful for a
man to fet his feet upon, when he dives; but efpecially
for finking of the *Ark*.　For this being made of Timber;
and full of Air, cannot of 'its own accord go down, unlefs
it be pulled, and forced by fome weight.　It may either
be broad and round, or fquare: if fquare, a large foot over
from fide to fide, or 16 inches will determine the breadth.
By this means, it will happen to be pretty thick, feing a
great quantity of Lead is required.　In each corner, there
muft be a hole, for four chords, by which it is appended to
the mouth of the *Ark*.　Between it, and the roof within,
muft be the height of a man and more.　The weight of
it, cannot be well determined without trial; feing it
depends upon the dimenfions of the Ark.　Firft then try,
how much weight,will bring the top E F G H level with
the furface of the Water.　When this is found, add a
little more weight till it begin to fink, and this will furely
take it to the ground,though it were 40 fathom.　'Tis to
be obferved, that when the top E F is level with the furface, there is here a juft counterpoife, namely between the
Lead foot-ftool on the one part, as a *pondus*, and the *Ark*
on the other part, as a *potentia*; for with what force the
Ark endeavours to pull up the *Lead*; with the fame force
ftrives the *Lead* to pull down the *Ark*.　Hence it is, that
as a fmall weight will turn a pair of Scales, when they
are in *equilibrio*; fo a fmall weight added to the *foot-ftool*
will fink the *Ark*.　Though it may feem difficult to
determine the juft weight of the foot-ftool, without trial

as I said, yet I purpose to essay it. For this cause consider that there is no Vessel of VVood almost, if it be once
full of Water, but the orifice of it will ly level with the
surface of the VVater, wherein it sweems. This proposition is so evident from experience, that it needs no confirmation. From this I gather, that as much weight of
Lead or Stone will bring the top of the Ark E F G H,
level with the surface of the VVater, as is the weight of
the Water, that fills it. If you suppose then the Ark
to be 36 inches broad, and 40 inches high, it must contain 30 cubique foot of Water. Now, supposing each
square foot of this Water to weigh 56 pound, 30 foot
must weigh 1680 pound. This is gathered from trial
and experience, for after exact search, I found a cubique
foot of Water, in bulk about 16 pints of our measure, to
weigh 56 pound. Take then a piece of Lead of that
weight, and you will find it make a just counterpoise with
the Ark. If any be desirous to know the quantity of it,
I answer, if lead be 13 times naturally heavier then Water,
you will find that a piece of Lead about 16 inches every
way will do it. If it be objected, that when a mans body
is within the Ark, the weight of the foot-stool must be
less, even as much less, as is the weight of the man, whom
I suppose to weigh 224 pound, or 14 stone. I answer,
the whole weight of the man is not to be deduced from
the foot-stool, but the one half only, and the reason is,
because a mans body being of the same specifick and natural weight with Water, it cannot preponderat or weigh
in VVater, because magnitudes only naturally heavier
then VVater weigh in VVater, as Lead, or Stone;
therefore seing the one half of the man is within the Ark,
and the other without among the Water, that part only

must

[Marginalia, handwritten:]
1. Foot
Water weigh
56 ℔

13 Water
= 1 Lead

16 = 1 Stone
14
64
16
224 ℔

muſt weigh, that's invironed with Air. This may ſeem a plauſible anſwer, and might do much to ſatisfy theſe, that are not very inquiſitive, yet, being examined, it will be found unſufficient. Therefore, I ſay, there's not one part of the mans body, that weighs within the Ark, or makes it heavier. Yet, I affirm, that when the mans body is within the Ark, a leſs weight will ſink it, then when his body is out of it, even as much leſs than before, as is the juſt weight of the one half of the man. For example, if 1680 pound be the juſt counterpoiſe of it without the Man, then after the Man is in it, it will take only 1568 pound to counterballance it, ſuppoſing the one half of the man to weigh 112 pound, or ſeven ſtone: yet it is not the weight of the man that makes this difference. For underſtanding what's the cauſe of this alteration, con-ſider, that when a mans body is within the Ark, there is leſs Air in it, then while his body is out of it, even as much leſs in quantity, as the bulk of the parts are, that are within. If this be, then muſt the Ark become heavier, not becauſe the mans body makes it heavier, but becauſe there is leſs Air, in the Ark, then before, and therefore, there ariſes an inequality between the weight of the foot-ſtool and the weight, or rather lightneſs of the Ark. For if 1680 pound of Lead, was the juſt counterballance of it, when it had 30 cubique foot of Air within it, it muſt exceed, when there is leſs Air in it. But there occures, here two difficulties, the firſt is, what's the reaſon, why as much weight muſt be deduced from the foot-ſtool, as is the the preciſe weight of the one half of the man ? Secondly, how ſhall we come to the true knowledge of that weight; that is, to know diſtinctly how many pounds or ounces it is of ? For anſwer, let us ſuppoſe, that the one half of

the

the man, is juſt as heavy, as ſo much Water equal in bulk
to his own half. This may be granted without ſcruple,
ſeing a mans body is judged to be of the ſame ſpecifick,
and natural weight with Water: and though there ſhould
be ſome ſmall difference, yet it will not make, or produce
any inſufficiency in the argument, for theſe demonſtrations, are not Mathematical but Phyſical. Therefore, as
much Water in bulk, as is equal to that part of the man,
that is within the Ark, muſt be as heavy, as the half of
the man. Now ſuppoſing the half of the man, to weigh
112 pound, and conſequently that Water, to weigh
as much, I affirm the ſaid Water to contain 3 4 5 6
cubique inches: but 3456 cubique inches, makes exactly
two cubique feet, which I gather thus. Seven pound of
Water requires 216 cubique inches, becauſe a Cube of
ſix inches, weighs exactly ſeven pound, therefore according to the rule of proportion, 112 pound will require
3 4 5 6 inches, which amounts to two cubique foot. The
Ark then by receiving the one half of the mans body,
loſeth two cubique foot of Air, therefore if 30 foot of Air,
require 1680 pound weight of Lead to counterpoiſe it,
28 foot of Air, muſt require only 1568 pound: therefore
to make a new counterballance, you muſt deduce 112
pound from the foot-ſtool. This anſwers both the difficulties. If it be ſaid, that the foot-ſtool weighs leſs in
VVater than in Air, therefore it muſt be heavier, then
1680 pound. I anſwer, 'tis needful to abſtract from that
difference, till the juſt calculation be once made, and
that being now done, I ſay, that a Cube of Lead 16
inches weighing 1680 pound, (If Lead be 13 times
heavier than VVater,) will loſe about 130 pound. The
reaſon is evident, becauſe a heavy body weighs as much

X

leſs

less in VVater than in Air, as is the weight of the
Water it expells. But so it is, that a Cube of Lead
of 16 inches expells a Cube of VVater 16 inches:
But a Cube of VVater 16 inches weighs 130 pound,
which I gather thus. 216 inches, or a Cube of six inches,
weighs seven pound, therefore 4032 inches, must weigh
130 pound. For if 216 give 7, 4032 must give 130. But
to return. Though there be small difficulty to let it
down and to sink it 20 or 30 fathom, yet there is no small
difficulty to pull it up again. And the reason is this,
because the further down it goes, the Air within, is the
more contracted, and thrust up, by the Pressure of the
Water, towards the roof. By this means, though near the
top of the Water, there was little difference between the
weight of the *Lead* and the *Ark*; yet 9 or 10 fathom down,
the difference is great, the weight of the one, far exceeding
the weight of the other, and therefore there must be greater difficulty to pull it up from 10 fathom, than from 5: and
yet more difficulty from 20 than from 10. However,
yet 'tis observable that, as the *Ark* in going down, becomes heavier and heavier, so in coming up, it growes
lighter and lighter: therefore less strength is requi ed, in
pulling it up from the tenth to the fifth fathom, than
from the fifteenth, to the tenth: the reason is, because
in coming up, the Air within expands it self, and fills
more space in the Ark, which in effect makes it lighter,
and more able to overcome the weight of the *Lead*. To
make these things more evident, let us suppose, that when
the *Ark* is down 18 or 20 fathom, the Air to be contracted
by the force of the Water, from L M to P Q 12 inches.
Next, that the weight of the *foot-stool* is 1680 pound.
Now, if this weight was the just counterpoise of the *Ark*,

at

at the top of the Water, then surely it must far exceed
it now, when it's 20 fathom down, becaufe the Air that
was 30 foot, is now reduced to 21. Count then, and you
will find, that if 30 require 1680, 21 will only require
1176: therefore the weight of the *Lead*, will exceed the
weight of the *Ark*, at 20 fathom deep, by 504 pound.
This will be yet more evident, if we confider, that while
the top of the Ark E F G H, is level with the furface
above, the VVater thruft out of 'its own place by this
bulk, is juft the weight of both *Lead* and *Ark*. But
when 'its down 20 fathom, and the Air reduced from L M
to P Q, there cannot be fo much VVater expelled now
as before, feing the fpace L M P Q is full of VVater.
Now, I fay, the *Lead* at 20 fathom, muft be exactly fo
much heavier than the *Ark*, as is the weight of the faid
VVater L M P Q, which in effect will be 504. pound:
for 'its a fquare body, 36 inches in thickneſs and 12 in
deepneſs. The weight of the rope is likewife to be
confidered, that lets down the *Ark*: for the longer it be,
and more of it goes out, it's the heavier, and more trouble-
fome to pull up.

There is no way to cure this difficulty, but by finding
out a way, how to keep a juft counterpoife between the
Lead and the Ark, all the time it is in going down. If
the Air within did not contract it felf, no difference would
happen: but this is impoffible, fo long as the Water is
under a Preffure. The expedient then muft be found out
another way, namely by knitting a fmall rope to the iron
ring N, in length with the other, to which at certain di-
ftances, relating to the fathoms the *Ark* goes down, muft
be faftned empty little Veffels of Wood, or bladders,
which by their lightneſs, may compenfe the decrement

and

and decreasing of the Air. First then, let down the *Ark* three fathom, and see how much it is heavier than before : and as you find the difference, so fasten to R one Bladder, or two, till the *Ark* be brought near to a counterpoise. Secondly, let it go down other three fathom, and obferve that difference alfo, and accordingly fasten to T as many, as will reduce the two to a counterpoise again. Do after this manner, till it sink 15 or 20 fathom. 'Tis to be obferved, that the further down the Ark goes, the difference is the lefs : therefore lefs addition will ferve : and the reafon is, becaufe there is lefs Air contracted, in paffing between the fifth and the tenth fathom ; than in paffing from the fiift to the fifth. The proportion of contraction is reprefented by the unequal divifions within the mouth of the Ark, as 1. 2. 3. 4. In a word, by what proportion the decrement of the Air is, by that fame proportion muft the addition be, upon the rope S N. Suppofe then, the Air to be diminifhed four inches, in going down four fathom, which will be 5184 fquare inches, or three fquare foot, then furely as much Air muft be added to the rope S N, by bladders. In going down as far, let us fuppofe three inches to be contracted; then lefs will fuffice. Though it cannot be determined without trial, how much Air is contracted in three fathom, and how much in fix, and how much in nine ; yet this is fure, that the decreafing is according to unequal divifions, that's to fay, lefs in fix than in four, lefs in 8, than in fix, and lefs in 10, than in 8, and fo downward : and that this is the rule, namely according to what quantity, the Air within the Ark is contracted, according to that fame meafure, muft the addition of Air be to the rope. If it be faid, that Bladders full of wind, cannot go down thorow the VVater without burft-

ing

ing. I anſwer, 'tis a miſtake, becauſe their ſides being
pliable, and not ſtiff like the ſides of a Timber Veſſel,
they yeeld, and therfore cannot burſt. It's obſervable
that when a bladder goes far down, the ſides becomes
flaccid and flagging. In this caſe, the Air, that before,
had the forme of the Bladder, and was ſomewhat ovall,
muſt now become perfectly globular, and round: for 'tis
ſure, that the dimenſions of it are altered by the Preſſure
of the VVater, namely from more quantity to leſs: if this
be, then the form muſt be round, ſeing the Preſſure of the
Water is moſt uniform; even as drops of VVater, or
Rain from a houſe ſide are round upon this account. This
ſecond way, may be thought upon alſo. Make the *Leaden*
foot-ſtool that ſinks the *Ark,* not of one piece, but of many,
that ſo, when the Air within it, begins to be contracted
by degrees, in going down, a proportionable weight may
be ſubtracted, for keeping a juſt counterpoiſe, all the
while of the deſcent. Or becauſe the greateſt trouble is
in bringing of it up, let the *Diver,* when once he is at the
bottom, ſubtract ſo much weight from the foot-ſtool, as
he thinks will go near to make a counterpoiſe, at that
deepneſs. For example, if the weight of the foot-ſtool
be 40 pound heavier than the Ark, then let him ſubtract
30 or 36, which may ly, and reſt upon the ground, till
it be drawen up, at a convenient time, by a chord. By
his means it will be eaſie to move the *Ark,* from one place
to another. Next, there ſhall be little or no difficulty to
pull it up. Nay, upon ſuppoſition, the rope were broken,
by which it was let down, yet if the *Diver* pleaſe, he may
come up without any mans help. And this is moſt eaſily
done, namely by ſubtracting as much weight, as will make
the Ark the ſtronger party. 'Tis to be obſerved, that
when

when you are at the bottom, and if you make the *Lead* but
one pound lighter than the *Ark*, it will surely come up,
and cannot stop by the way. The reason is, because a
very small weight will turn the Scales, between two bodies,
thus weighing in VVater. Next, the further the *Ark*
comes up, it becomes the lighter, because the Air within
it, expands it self the more. But leaving this, let us come
to explicat the reason, why the contraction of the Air
is not uniform, but rather difform. For if in going down
three fathom, three inches be contracted, there will not
be other three contracted in going down the second three,
but less: and yet less in going down the third three. Two
things then are to be explicated here. First, why there
is a contraction. Next, why it is after such a manner.
As for the first; the contraction is caused by the Pressure
of the Water, which gradually increaseth from the top to
the bottom; as is clear from the last Experiment: therefore, there being a greater Pressure in a surface six
fathom deep, than in a surface three fathom deep, the Air
within the *Ark*, must be more contracted in passing
between the third and sixth, than in passing between the
first and third. When I say more contracted, the meaning
is, that more quantity is contracted to less, whereby the *Bensil* of it is more intended; or that the Air is more bended.
As for the second, we must remember from the last Experiment, that the cause of this, is not from the VVater,
as if forsooth the Pressure of it, were according to unequal
proportion, but from the Air it self, whose kind and
nature it is, to suffer compression after such a way. 'Tis
evident in *Wind-guns*, whose second span of Air is comprest with greater difficulty, than the first: and the third
with greater difficulty, than the second. 'Tis so with
all

all bodies endowed w th Bensil : for ay the longer you bend , you find the greater difficulty. As there is a great disadvantage to the man that *Dives* , from the contraction of the Air, so there is a great advantage to him, from this manner and way of contraction, for if it were uniform, according to the Pressure of the Water , then if three fathom comprest three inches, six fathom ought to compresse six inches, nine fathom nine inches , and so forward, till by going down, either the whole Air , should be comprest to no inches , or else very little should remain for respiration.

The next thing to be taken notice of, is that all the while, during the down going of the *Ark*, there is still equality of weight, between the *Pondus* of the Water , and the *Potentia* of the Air , for with what degree of weight, the Water presseth up the Air, with the same degree of force and power, doeth the Air press down the Water. If this were not, it would be impossible for a man to go down, because of pain. For when one part of a mans body, is less prest than another , there ariseth a considerable pain, which sometimes is intolerable, as is evident from the application of *Ventoso-glasses*. This equality of weight, is the true reason , why respiration is so easie. Yet 'tis to be observed, that a man cannot breath so easily in the *Ark*, under the Water, as above in the Air ; not because there is any inequality, between the weight of the VVater , and the force of the Air ; but only because the quantity of it is little. For when a man sucks in as much Air, as fills his lungs, the quantity must be diminished : if this be, the Water must ascend by proportion, though insensibly. When a man thrusts out the same Air again, the quantity is increased ; if this be, then the Water must subside a little, both which cannot be

with.

without difficulty, ſeing there is a ſort of ebbing and flowing both of the Air and of the Water, in every reſpiration. But it rather ſeems (you ſay) that this difficulty flowes from the ſtrong, extraordinary benſil , that the Air is under.   I anſwer, as long as the preſſure of a Fluid is uniform, though in a high degree, yet there can be no trouble in reſpiration; becauſe with what force ſoever, it is driven in upon the lungs, with the ſame force it is driven out again: therefore, though the Air we live in, were as much again bended as it is, yet (as is probable) we would find no more difficulty in breathing than now. There is one thing makes breathing eaſie under the Water, in the Ark, namely this; when a man ſucks in the Air to his lungs, his breaſt and belly goes out, and ſo fills the ſpace deſerted by the Air , that goes in. This makes the ebbing and flowing far leſs.

From this equality of weight between the preſſure of the Water , and the preſſure of the Air , we ſee good ground to ſay , that though the *Ark*, were no thicker in the ſides, than a thin ſawed dale , yet there would be no hazard of breaking. I am confident, though it were no ſtronger in the ſides , than a wine-glaſs, that's ſoon broken; yet it might go down 40 fathom without hazard , or danger of burſting.   This affords good ground likewiſe to make windows in the Ark covered with glaſs : for if the Preſſure be uniform , and equal, its impoſſible they can be broken. The Water cannot thruſt them inward, becauſe the Preſſure of the Air, is as able to thruſt them outward.

It's certain, the more Air be in the Ark, the more eaſie is reſpiration : therefore its more eaſie to breath, when the Ark is but down 5 fathom, than when it is down 10 or 15.   It's probable a man might live within the *Ark*, it being 40 inches deep, and 36 inches wide, at the deepneſs of

ten fathom, near two houres ; whereas if it were round, and
narrow above in form of a Bell , he could not continue an
hour. It were very easie to try how long other creatures
might live in it, for example dogs, and such like, or fowls, as
hens, pheafants or doves. They might easily be incloted from
coming out; for though the whole mouth of the Ark were
fhut up, except as much paffage , as would receive a mans
fift, yet it will operate, as well that way, as the other. And
there, a little door might be made to open, and fhut at plea-
fure. 'Tis obferved, that by long tarrying under the Water in the *Bell*, the Air becomes grofs and mifty , which
hinders a man from feing about him. The caufe of this,
are vapors that come from the ftomach , lungs and other
parts of the body , efpecially from the ftomach , when the
ventricle is full of meat. It's not fit then, that a man about
to *dive*, fhould eat too much, or drink too much, efpecially
fuch liquors as *Sack* or *Brandy*, that beget many fumes and
vapors. If a man were neceffitated to tarry a pretty while
below, frefh Air might be fent down from above, in bottles
or bladders, even as much as might fill up the place deferted
by the contracted Air. 'Tis obferved by fome, that have
been under the VVater, that their eares have been fo troubled , that for a long time, they have found difficulty to
hear diftinctly. The reafon of this muft be from the great
Preffure , the *tympanum* hath fuffered from the imprifoned
Air of the *Bell*. The Organ of hearing is foon troubled,
efpecially when a man is near to a great *gun*, when it's fired.
And furely, when a man is but 34 foot down, the Air within the Ark, will be of double Benfil : put the cafe the man
go down 68 foot, or 13 or 14 fathom , the Benfil is tripled : that's to fay , if the Air above have five degrees of
Preffure in it, the Air of the *Bell*, at 68 foot deep , will

Y   hae

have 15 degrees of Preſſure, therefore the *tympanum* of the
ear that's but a ſmall and thin *membran*, muſt be ſore
diſtreſſed ; that is overbended , and preſt inward ; even
as , while a man ſets upon a drum head a great weight,
v. g. a Bullet of Lead or Iron, of 20 or 30 pound, the skin
by this, ſuffers an extraordinary Preſſure, whereby it is in
hazard to be rent. 'Tis probable, if a man ſhould go very
far down, the *tympanum* might be in hazard of breaking, or
being rent in two pieces, there being a greater Preſſure
upon the one ſide from the Air without , than upon the
other ſide, from the internal Air within, which is thought
to be within the *tympanum*.

There remains another *Phenomenon* to be explicated , and
it's this : the further up the *Ark* comes from the ground of
the Water, towards the top, the Water within it, ſubſides
and ſettles down more and more, towards the mouth. The
reaſon of it is, becauſe the further up , the Preſſure of the
Water is the leſs ; and therefore the contracted Air gets
liberty to expand, and dilate it ſelf, and ſo thruſts down the
Water from P Q to L M. In a word , by what proportion the Air is contracted in going down, by that ſame proportion it dilates , and opens it ſelf in coming up. This
lets us ſee , as there is diſadvantage in going down , from
the contraction of the Air , ſo there is advantage in coming up , from the dilatation of it. Some think, that the
coldneſs of the Water is the cauſe , why the Air is contracted in the Ark, ſuch are thoſe, who deny the Preſſure
of it. But this fancy is eaſily refuted ; becauſe in aſſert-
ing this , they muſt maintain, the further down , the cold
is the greater. If this be , then far more Air muſt be contracted, in going down from 10 to 15 fathom , than in
paſſing from 5 to 10 ; ſeing as they ſay , the further down,

the

the cold is the greater ; and therefore the contraction of
the Air mu t be the greater ; that's to say, there muſt be
more quantity of Air contracted in the one ſpace , than in
the other.   But ſo it is , that the further down , the contraction is the leſs.   They judge likewiſe the coldneſs of
the Water to be the cauſe , why the ſides of empty Vef-
ſels are broken in going down.   But if this be , then a
ſtrong Veſſel ſhould go no further down than a weak Vef-
ſel ; ſeing cold can pierce thorow the ſides of the one, as
well as thorow the ſides of the other.   And why is it, that
a bladder full of wind will go down 40 or 50 fathom without burſting , yea 100 , and yet a ſtone-bottle or glaſs-
bottle, cannot go beyond 20 or 30 ? If cold have in it,
that power to break the ſides of a ſtrong bottle, it muſt be
far more able to burſt the ſides of a thin Bladder. This difference is clearly explicated from the Preſſure of the Water, but I defy any man to ſhew the difference from the coldneſs of it.   'Tis to be obſerved , that in all ſuch Experiments of ſinking of Veſſels, as *Hogſ-heads*, *Barrels*, and
Bottles, they muſt be cloſs on all ſides.   Therefore, if a man
deſire to know , how far down a Glaſs-bottle is able to go
without burſting, he muſt ſtop the mouth of it exactly,
with a piece of wood, and cement.

In ſetting down the dimenſions of the *Ark* , I have re-
ſtricted them to 40 inches high , and 36 inches wide.
But if any man be deſirous to enlarge them, or make them
leſs , he may do it. Only 'tis to be obſerved , that the
larger the *Ark* be, the *Foot-ſtool* that ſinks it, muſt be the
heavier. Yet it hath this advantage, that it contains much
Air , which is the great perfection of it. One of a leſſer
ſize hath this advantage, that it's more tractable, and eaſier to let down, and to be pull'd up.   But theſe things are

best known from Experience, or if a man please, he may calculate.

As the *Ark* is a most useful device for profit, so 'tis excellent for pleasure, and recreation, if a man were disposed to see the ground and channels of deep VVaters, or were inclined to find out *Hydrostatical* conclusions, a knowledge very profitable, and which few have attained to. Though it seem somewhat difficult to enter the *Ark*, and go down below the Water, yet a little use will expell all fear. Then, a man may go down with less hazard, and fear in the *Ark*, then in the *Bell*, because he may conveniently fasten his hands, to each side of the Ark, if need were. He may conveniently sit, as in a Chair, all the time of down going, and up-coming, by fixing a little seat in it: he may have windows to look out at: his body may be so fixed, that there needs be no fear of falling out.

If a man were desirous to make *Hydrostatical* conclusions, by *Diving* under the VVater, the dimensions of the *Ark* might be enlarged, so that it might conveniently cover a mans whole body, by which means, having much Air in it, a *Diver* might continue under Water half a day, if need were. Let us suppose then, the hight of it to be 8 foot, and the breadth 3 foot, or more. In such a case, a man might continue under the VVater many hours; and yet not one part of his body wet: for if the *Ark* be 8 foot high, and the man 5 foot in stature, at the deepness of 10 fathom, the Water can scarce rise 3 foot in it. But why may not a man come up every half hour, when he finds difficulty to tarry down in a little *Ark*? I answer, he may; but it's trouble and pains to pull him up, and let him down so frequently. And it may so happen, that through want of Air in a small *Ark*, he be necessitated to come up

before

before he end his work. And leaving the work imperfect, he may find difficulty in the second down going, to find sometimes the place where he was, or the thing he was about to lift, *v. g.* a chest of Gold. If it be said, that a great weight of Stone or Lead is required to sink an Ark 8 foot high, which will amount to 4032 pound weight. I answer, 'tis so indeed: but here is the advantage; when it is once below the Surface, there's little more trouble, then with an Ark of lesser dimensions; because of the *equipondium* that's between it, and the weight, that sinks it.

In such a Vessel many trials might be made. As first, that of the *Torricellian-Experiment*, which is nothing else, but a Glass-Tub so many inches long, with a *Mercurial* Cylinder in it of 29 inches high, that's supposed to be kept up at that hight by the Pressure of the Air. If this were taken down about 34 foot, 'tis very probable the Mercury would rise other 29 inches. The reason is, because the Air within the Ark, that presseth upon the Surface of the stagnant Mercury, must be under as much pressure again, as the Air above; but the Air above, is able to support 29; therefore this Air must sustain 58. The reason why the Bensil is exactly doubled is this, 34 foot of Water hath exactly as much Pressure in it, as the whole element of Air; therefore, the Air within the Ark, being 34 foot down, must not only have in it the Pressure of the Air above, but the Pressure of the Water likewise: this necessarily follows, because when two Fluids touch, or are contiguous to other, the one cannot be under five degrees of Pressure, unless the other be under as many. According to this reasoning, if the *Ark* go down 68 foot, the Mercury will rise from 58 to 87. If to 102; it rises

116.

116. This reckoning is founded upon this, namely that Water is 14 times lighter than Mercury; and therefore one inch of Mercury requires 14 of Water to support it in a Tub, and therefore, before Water is able to raise 29 inches of it, the Pipe muft be 34 foot deep.

For a fecond trial, blow a Bladder as full of wind as it can hold, and having knit the neck about with a Packthreed, place it in the *Ark*, and you will find the fides, that hath been ftifly bended become flaccid and feeble, as if the one half of the Wind had gone out, and this will come to pafs, before the *Ark* can go down eight or nine fathom. The ftrong benfil of the Air within the Ark is the caufe of this : for as the Ark goes down, the Air grows ftronger, and fo at length becomes of that power and force, that it eafily overcomes the force and Benfil of the Air of the Bladder, and reducing it to lefs room, caufes the fides become flagging. In this cafe, the faid Air, that was oval, and had the form of the Bladder, muft become round in form of a Globe, becaufe of the uniform Preffure, that it fuffers from the Air of the *Ark*. When once the *Ark* is down 14 or 15 fathom, take the fame bladder, and blow it ftiff with Wind, and knit the neck as afore. And you will find that in the up-coming, the fides of it will burft afunder with a noife. When the Bladder is thus full of Wind, 'tis fuppofed, that there is a fort of counterpoife between it, and the Air of the *Ark*. But as the *Ark* afcends, the Air of it, becomes weaker and weaker, while in the mean time, the Air of the Bladder fuffers no relaxation ; therefore, when the *Ark* comes near the furface, there arifes a great difproportion between the one Air and the other, as to ftrength, and therefore the Air of the Bladder being the ftrongeft, rents the fides

in

in pieces, and comes out with a noiſe. Or, blow it
but half full of wind, and you will find before, the Ark
come near to the top, the ſaid Bladder to be bended to
the full.

For a third trial, take a Glaſs, ſuch as they uſe in
Caves, for preſerving of Brandy, and ſtopping the mouth
cloſely, take it down with you in the *Ark*; and you will
ſee, the ſides of it break in pieces, before you go down
four or five fathom. The ſtrong Benſil of the ambient
Air, is the cauſe of this. If you take it down with the
orifice open, no hurt ſhall befal it. Or if you ſtop the
orifice in the up-coming, you will find the ſame hurt
come to it. But here is the difference, in the firſt burſting, the ſides are preſt inward, by the ambient Air; in
the ſecond, the ſides are preſt outward, by the Air within the Glaſs.

For a fourth trial, take a round Glaſs-bottle, pretty
ſtrong in the ſides, and when it is down with you in the
Ark 14 or 15 fathom, ſtop the mouth of it exactly, and
when it comes above, you will find a conſiderable quan-
tity of Wind come out of it, when the orifice is opened.
This evidently demonſtrats, that the Air within the
Ark, 12, 13, or 14 fathom down, is under a far ſtronger
Benſil then the Air above.

For a fifth trial, let a man apply to his skin a cold
Cupping-Glaſs, when he enters the Ark; and he will
find ſuch a ſwelling ariſe within it, as when it is applied
hot by a Chyrurgion. This tumor begins to riſe, aſſoon
as the *Ark* begins to go down. The reaſon is evident
from unequal Preſſure; the parts within the Glaſs being
leſs preſt, than the parts without.

For a ſixth trial, take a common *Weather-Glaſs*, and
place

Place it in the *Ark*, and in the going down, you will see
the liquor creep up in it, by degrees, as the *Ark* goes
down, as if some extraordinary cold, were the cause of
it. And as the *Ark* comes up by degrees, the said
liquor creeps down by degrees. The cause of this *Phenomenon* is not cold, as some might judge, but the strong
Bensil of the Air within the Ark, that so presseth upon
the surface of the stagnant Water, that it drives it up.
If you take with you, a *Weather-Glass*, *hermetically*
sealled, no such thing will follow; because the outward
Pressure is keeped off. 'Tis not then cold, that's the
cause, but weight. By the way take notice, that all
common *Weather-Glases* are fallacious and deceitful;
because the motion of the Water in them, is not only
caused by heat, but by the weight of the Air, which
sometimes is more, and sometimes less, as frequently I
have observed, and as hath been observed by others.
This difference is found, by the alteration of the altitude of
the Mercurial cylinder, in the *Baroscope*, which is more and
less, as the Pressure of the Air changeth. In fair weather,
and before it comes, the Mercury creeps up. In foul and
rainy weather, and a pretty while, before it fall out, it
creeps down. Because in fair weather, the weight of
the Air is more, than in rainy and dirty weather. *December*, 13. 1669. I found the altitude 29 inches, and nine
ten parts of an inch: at this time the heavens were covered with dry and thick clouds, and no rain followed.
March 26. 1670. I found the altitude no more, than 27
inches, and nine ten parts, at which time, there was a
strong Wind with rain. Between these two termes of
altitude, I have found the Mercury move near a twelve
moneth. 'Tis a most sure prognosticator, for if after
rain,

rain, you find the Mercury creep up in the morning, you may be ſure, all the day following will be fair, notwith-ſtanding that the heavens threateneth otherwayes. If after fair weather, the Mercury ſubſide, and fall down a little, you may be ſure of rain within a ſhort time, though no appearance be, in the preſent. It falls down likewiſe, when winds do blow. What the true cauſe is, why there is ſuch an alteration in the Preſſure of the Air, before foul weather, and fair, and in the time of it, it is not eaſie to determine.

But we proceed. Trial likewiſe might be made, by firing a great piece of Ordnance above, whether the re-port would be heard below the Water or not? This would determine the queſtion, whether Water be a fit *medium* for conveying ſound as Air is. Item, whether or not, the Sea water be freſher at the bottom, than near the top, which is affirmed by ſome. Item, whether ſounds be as diſtinct in ſuch a ſmall portion of Air, as they are above. This might be tried with a Bell of a Watch. If need were, a little chamber Bell might be hung within the *Ark*, and a ſmall chord might paſs up from it, through the cover, whereby the perſons above, might by ſo many tingles, ſpeak ſuch and ſuch words to the *Diver*. I have demonſtrated before, that though there were a little nar-row hole made in the cover above, yet neither Air would go out, nor Water come in. If a man were curious, he might have a window not only in the ſides, but in the roof above, covered with a piece of pure thin Glaſs, thorow which he might look up, after he is down two or three fa-thom, and ſee whether there appeared any alteration in the dimenſions of the body of Sun or not, or ſeemed nearer.

Z   VVe

We now come to infer some *Hydrostatical* conclusions, as from former Experiments. We see then first, that in Water there is a pressure; namely from the strong *Bensil* of the Air within the *Ark*, that groweth stronger, and stronger, as the Water groweth deeper, and deeper. We see next, that the pressure of the VVater hath an increment : because the further down the *Ark* goeth, the Air is the more bended. Thirdly, two Fluids cannot be contiguous one to another, unless both of them be under the same degree of pressure : because the Air of the *Ark*, and the Water that creepeth up within the mouth of it, are perpetually under the same degree of power, and force, whatever the deepness be. Fourthly, that in Fluids the pressure is uniform; because the Air of the *Ark*, and the Water without, press most equally, one against the other. Fifthly, the more that the Air is bended, it is the more difficult to bend it; and consequently, that the diminution of the quantity, is according to *unequal proportion*. Sixthly, that when the *Ark* is down 34 foot, the *Bensil* of the Air is doubled : and tripled, when its down 68 foot : because the pressure of 34 foot of VVater, is as much as the whole pressure, that's from the *Atmosphere*. If it be enquired, how much weight rests upon the palm of a mans hand, when the *Ark* is down about 68 foot? I answer, the pressure of the Water upon a mans hand, at that deepness with the pressure of the Air above, will be equivalent to the weight of a pillar of Mercury 87 inches high, and three inches thick, which will exceed in real weight 200 pound. If so much rest upon the palm, how much must rest upon the rest of the parts of the body? Let us suppose then, the quantity of the palm, to be found in a mans skin, 200 times, then must he suffer as much

Pressure,

preſſure, and actually ſupport as much burden, as will amount to 40000 pound weight. Seventhly, our bodies may be under a huge preſſure, and yet that burden not perceptible ; as is evident from the *Diver*, who findeth little or no weight, while he is under the Water. Or if there be any Preſſure found, it's not comparable to that, which really is, Eighthly, when a man is 14 or 15 fathom down, at every inſpiration and expiration, his breaſt and belly muſt lift up the weight of 1800 pound : becauſe, if the whole burden be 40000, the weight that reſts upon the breaſt, and belly, will be about 1800. Ninthly, that between every inſpiration, and expiration, there happens a perfect counterpoiſe, namely by the Air, that goeth into the lungs, and the outward Air of the *Ark* : for if the Preſſure of the one, were more, than the Preſſure of the other, there could be no motion of the lungs. Tenthly, when a man draweth his breath, the Air cometh not in by *ſuction*, but by *pulſion*. For this cauſe, though the VVind-pipe were ſtopped, yet a man might live by having a hole in his ſide, going into the lungs. Laſtly, that there is no ſuch thing as *ſuction* properly ; and therefore the motion of all Fluid bodies, is cauſed by Preſſure and weight. The motion of the blood then thorow the heart, is *driven*, and not *ſucked*. Infants properly do not ſuck, but have the milk ſqueezed into their mouth. 'Tis evident from the ſucking-glaſs that ſome women uſe for milking their own breaſts : for by help of this, the Air that guardeth the head of the Pap is removed, and ſo the Air, that preſſeth the parts about, and without, ſqueezes out the milk.

Z 2 EXPE-

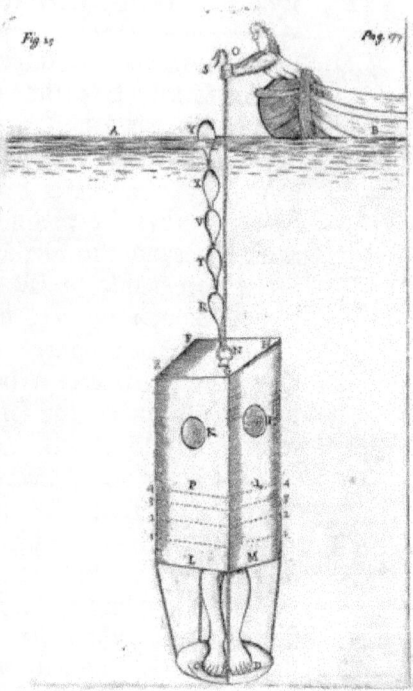

Fig 25.   *Pag. 177.*

EXPERIMENT XIX.
Figure 26.

THis Figure reprefents a deep Water, whofe firft and visible furface, is F G. The imaginary furface, is E L C, 34 foot below it. A D B is a *Siphon*, working below this VVater with Mercury. A E L is a Veffel with ftagnant Mercury, among which the orifice A is drowned, the other orifice B exifting among the Water, D M is the hight of the *Siphon* above the line of level, which I fuppofe is 58 inches. For making it work, ftop the two orifices clofely, and pour in as much Mercury at a hole made at D, as will fill both the legs. Then ftopping the faid hole, open the two orifices A and B, and you will find the liquor run as long out at B, as there is any almoft in the veffel A E L. For evincing this, which is the only difficulty, confider, that if this *Siphon*, were filled with Water, and made to work only with Air, (as is clear from daily experience) the liquor would run out conftantly at B. Becaufe there is here an unequal Preffure ; the furface of Air N B, being more burdened, than the furface E L C, but where unequal Preffure is in Fluids (according to the 12th Theorem) motion muft follow, I prove the furface N B to be more burdened, than the furface E L C, becaufe the Water B D, is heavier than the Water L D, as is evident to the eye. The Air B therefore, fuftaining far more weight, than the Air E L, muft cede and yeeld. Next, there is here a *pondus* and a *potentia*, the *pondus* is the VVater L D ; the *potentia* by which it is counterpoifed, is the Water B D ; but thefe are unequal, B D being

heavier,

heavier, than L D ; therefore according to the 33 Theorem, these two Fluids cannot cease from motion. If it be said, that the surface N B is stronger, than the surface E L C, seing it is lower. I answer, the difference is so unsensible, that they may be judged but one. Now, I say, if this *Siphon* work in Air, with Water, it must likewise, work in Water with Mercury. Therefore, this *Siphon* being 34 foot below the first surface F G, the liquor must run out constantly at B. Because, there is here, an unequal Pressure, the surface of VVater N B, being more burdened, than the surface E L C. Though there be more weight in N B, than in E L C, because it is lower, yet because the difference is not so much, as is between the weight of B D, and the weight of L D, it proves nothing. Note here, that so long as D, is within 58 inches of E L C, this *Siphon* will work. The reason is, because the Pressure of 34 foot of VVater; with the Pressure of the Air, upon F G, are able to raise Mercury exactly 58 inches. But if D exceed that hight, no Art will make the liquor run out at B. Note secondly, that this *Siphon* will operate with Air and VVater, though the top D were 34 foot above M ; and the reason is, because the Pressure of the Air, is able to raise a pillar of Water to that hight. Note thirdly, that if there were an orifice opened at C, upon the level line E L C, the two Waters would become of the same weight, the one not being able to move the other. If you bore a hole at R, the liquor ascends from R to D, and goeth down from D to A, and so the motion ends. But, if the leg A D were six times wider, than B D, the liquor would not run out at B. I shall answer this in the close.

From this Experiment we see first, that the motion of

Fluid Bodies up thorow *Pumps* , and *siphons* is not for
fhuning *vacuity*, but becaufe they are preft up violently.
We fee next , that when the Preffure is uniform , there
is no motion in Fluids; but affoon , as one part is more
preft, than another, motion begins : becaufe, this *Siphon*
will not operate,if the orifice be made in C ; but if fo be,it
be in D , then the motion begins; becaufe there is here
an unequal Preffure, which was not in the other. We
fee thirdly , that Fluids have a determinate *Sphere* of
activity, to which they are able to prefs , and no further :
becaufe this Water , is not able to prefs Mercury higher
than 58 inches. So the Air cannot raife Water higher
than 34 foot. If this Water were 68 foot deep , the
Sphere of it's *activity* would be 116 inches. We fee
fourthly , that in Fluids there is a *Pondus* and a *Potentia* ;
and that the inequality of weight between the two , is the
only caufe of motion. We fee fifthly , that as long as
this inequality of weight continues , as long continues the
motion , becaufe , as long as B D , is heavier than L D,
the motion perfeveres. We fee fixthly, the poffibility
of a *perpetual motion* in Fluids; becaufe the liquor runs
perpetually out at B. If it be faid , the motion ends,
when the ftagnant Mercury A E L faileth. I anfwer,
this ftop is only accidental , and not effentially from the
nature of Fluids. If it be enquired, whether or not,would
the Mercury run out at B , upon fuppofition , the fhank
L D were twice as wide, as the fhank B D? I anfwer
it would. If it be faid that the one is far heavier than the
other, namely L D than D B. I anfwer, weight in
Fluids is not counted according to thicknefs , but according to altitude.

EXPE-

EXPERIMENT XX.
Figure 27.

THis laſt is for demonſtrating the preciſe and juſt weight of any Pillar of Air, Water, Mercury, or of any other Fluid body, if ſome of their dimenſions, be but once knowen. A B then is a ſquare Pipe 12 foot high, and ſix inches in wideneſs, full of Water, reſting upon the ſurface of Air A C. And E G is a ſquare Pipe 12 foot high, and 12 inches wide, full of VVater, reſting upon the ſurface of Air E F. None needs to doubt, but the two Waters, will be ſuſpended after this manner, even though the orifices A and E were downward, eſpecially if they be guarded with Water, but the demonſtrations, will be the more evident, that wee ſuppoſe the two Pillars of Water to be ſuſpended as they are. From this Experiment I ſay firſt, that the Pillar of Air C D is 168 pound weight, at leaſt, which I prove thus. The VVater A B is 168 pound: therefore the Air C D, muſt be as much. I prove the *Antecedent*, becauſe it's a Pillar of VVater 12 foot high, and ſix inches thick: but every half cubical foot of VVater, that containes 216 inches, weighs ſeven pound: therefore ſeing the Pillar is 12 foot, it muſt contain 24 half feet; but 24 times 7 is 168. The only difficulty is to prove the *Connexion*, which I do thus, from the ſeventh Theor. *all the parts of a Fluid in the ſame Horizontal line, are equally preſt*, but ſo it is, that the part A, and the part C, are in the ſame horizontal ſurface; therefore the part A, and the part C, are equally preſt. But if the part A, and the part C, be equally preſt, the Pillar of Air C D, muſt be as heavy, as the

Pillar

Pillar of VVater A B. I fay fecondly, that the Pillar
of Air F H, weighs 672 pound, I prove it thus. The
Water E G weighs 672 pound; therefore the Air F H,
weighs as much. The *Antecedent* is clear, becaufe E G,
is a fquare Pillar of VVater 12 foot high, and 12 inches
thick ; but every cubical foot of VVater weighs 56
pound: but 12 times 56, is 672. I prove the *connexion*,
as before. All the parts of an horizontal furface, are
equally preft ; therefore the part F, muft fuftain as much
burden, as the part E.

To proceed a little further, let us fuppofe the Pipe
A B to be 34 foot high, and the Pipe E G to be as
much. I affert then thirdly, the Pillar of Air C D to
weigh 476 pound, which I prove as before. All the
parts of the fame furface, are burdened with the like
weight, but the part A fuftains 476 pound, therefore the
part C muft fupport as much. The *Connexion* is evident,
and the *Antecedent* is fo too, becaufe the VVater A B
being 34 foot high, and fix inches thick, muft weigh
476 pound: for, if 216 inches, weigh feven pound,
14688 inches, muft weigh 476 pound. I affert fourthly, the Pillar of Air F H to weigh 1904 pound, which
I demonftrat by the former *Medium*. All the parts of a
Fluid that ly in the fame horizontal furface, are equally
preft ; but fo it is, that E and F, do fo ly ; therefore F
muft be as much burdened as E; the Water therefore E G,
weighing 1904 pound, the Air F H, muft weigh as
much. For if 216 inches of Water weigh feven pound,
58752 inches (for fo many are in the Water E G) muft
weigh 1904 pound.

Let us fuppofe fecondly, the Tub A B to be only 29
inches high, and the Tub E G, of the fame hight, and
that

that six inches wide, and this 12 inches wide. I affirm
then fifthly, the Air C D to weigh yet 476 pound, and
the Air F H, to weigh 1904 pound. Because the Pillar
of Mercury A B, weighs 476 pound, and the Pillar
of Mercury F G, weighs 1904 pound: therefore, if A B
be 476, C D must be as much. And if E G be 1904,
F H, must be of the same weight. I prove the Mercury
A B to weigh about 476 pound, though it be but 29
inches high; because it is 14 times heavier then Water.
For the same cause, doth the Mercury E G weigh about
1904 pound. I say *about*, because 34 foot, containes 29
inches, more than 14 times.

Let it be supposed thirdly, the Pipe E G, (being 34
foot high,) to have the one half of it I G, full of Air, and
the other half E K full of VVater, I affirm then sixthly,
the part E, and the part F, to be yet equally burdened.
That's to say, the VVater E K, that's now but 17 foot,
makes as great a Pressure upon E, as when it was 34 foot.
The reason of this, is surely the Pressure of the Air I G,
that bears down the Water K E, with the weight of 952
pound, the half of 1904 pound. If it be said according
to the Theorem 21, that there is as much Pressure and
weight in the least part of a Fluid, as in the whole; therefore the Air I G, must be as heavy as E H. I answer
I G, is not so heavy as F H, because the Water E K
impending in the lower part of the Tub, hath occasioned
the Air I G, to expand it self so many inches, by which
means, it loseth so many degrees of it's *Bensil*. If you
remove the Water E K, then will the Air I G, be as
heavy, as F H; because E K being Air, it reduceth
I G to that same degree of *Bensil* with it self; but when
the Air E is burdened with the Water E K, it can-

not make the Air I G, of that fame weight with it felf.

Let us fuppofe fourthly, that only eight foot and an half of Water, are in the Tub, namely between E and N. I fay then feventhly, that the part E, is as much burdened with it, as when the Pipe was full; becaufe the 25 foot, and an half of Air N G, is exactly as heavy, as the 25 foot and an half of the Water that's gone. I prove it thus. The Air E hath the weight of 1904 pound in it felf, feing the weight of the furface, is alwayes equal to the weight of the Pillar, but being burdened with the VVater E N, that weighs 476 pound, it cannot prefs up with more weight then with 1428 pound: and therefore the top of the Water N, muft prefs upon the under part of the Air, that's contiguous with it, with 1428. If this be, the Air N G, muft prefs down with as much, feing according to the 20 Theorem, it is impoffible, that one part of a Fluid, can be under Preffure, unlefs the next adjacent part, be under the fame degree of Preffure. Therefore I conclude, that the 25 foot and an half of Air N G, is as heavy, as the 25 foot and an half of the Water that's gone. This makes it evident alfo, that when the Pipe is half full of VVater, as E K, the Air I G, hath the weight of 952 pound. Becaufe E being in it felf 1904, but being burdened with E K 952, it cannot make the top of the Water K, prefs upon I with more weight than 952; and therefore (by the 20 Theorem,) the Air G I, muft weigh 952 likewife.

I affirm eighthly, that, when the Pipe is full of Water, from E to G, if a man poife it in his hand, he doth not find the weight of the Water E G. And the reafon is, becaufe it's fuftained by the part of the furface E. But if the Air E fuftain it, my hand cannot fuftain it. I find
then

then only the weight of the Tub , but not the weight of
the VVater within it. I say ninthly, that when I poise
the said Tub, I find the whole weight of the Pillar of Air
L M, which is exactly 1904 pound. I prove it thus.
The *pondus* of a Fluid is then only found, when there is
not a *potentia* to counterpoise it, or at least, when the *potentia* is inferior to the *pondus :* but there is here no *potentia,* counterpoising the *pondus* of the Air L M. Therefore, I must find the weight of it, when I lift up the Tub.
The *major proposition* is clear from the tenth Theorem. It's
evident also, from common experience; for while a ballance is hanging upon a nail, with six pound in the one
scale, and nothing in the other, you will find the whole
burden, if you press up that one scale with the palm of
your hand. But if so be, there were six pound in the opposite scale, you will not find the first six ; and the reason
is, because it is in *equilibrio* with other six. 'Tis just so
here, I must find the weight of the Air L M , while I
poise the Tub, because it wants a weight to counterballance it. I prove the *minor proposition* thus. If any thing
counterballance the Air L M, it must either be the Air
below, namely the part E ; or the Water E G : but neither of the twain can do it. Not the Air E, because it
hath as great a burden upon it, as it is able to support,
namely the Water E G, that weighs 1904 pound. And
for this cause, not the VVater it self, seing all the force
it can have to counterballance L M, is from the surface of
Air E ; but this is in *equilibrio* with it already. I said
that the Air L M, was exactly 1904 pound weight. This
also is evident, because it is just of these same dimensions,
with the Air F H. If it be said, the Air L M must be thicker ; seing it's equal to the Tub without ; but the Air

F H, is only equal to the Tub within. I anfwer, it is fo indeed; but here is a folution to the difficulty. I do not find the whole weight of the Air L M, but only as much of it, as is equal to F H. Suppofe the Tub to be 12 inches within, from fide to fide, and 16 without; from fide to fide. I fay then, I find only the burden of fo much Air, as anfwers to the cavity of the Tub, becaufe the reft of thefe inches, are counterpoifed, by as much below, namely by the Air, that environs the orifice E: for it's fuppofed, that if the Tub be two inches thick above, it muft be as thick in the lips. So that the whole Tub, is not unequally preft, but only fo much of it within upon the top, as anfwers to the cavity. Tenthly, that when the Pipe is but half full of VVater, namely from E to K, I find only 952 pound of the Air L M, though before I found 1904. The reafon is, becaufe the one half of it is now counterpoifed by the Air I G, and therefore the weight of it becomes infenfible. 'Tis clear from the fixth affertion, that the Air I G, preffeth down with 952; therefore it muft prefs up with as much, feing according to the fixth Theorem, the Preffure of a Fluid is on every fide. Eleventhly, that when there is only eight foot of VVater and a half in the Tub, namely between E and N, I find only 476 pound of the Air L M. Becaufe in this cafe, the Air N G counterpoifeth 1428 pound of it: For if the faid Air, burden the Water N E, with 1428 pound, as is clear from the feventh affertion, it muft likewife prefs up the Tub with as much, and fo counterpoife as much of the Air L M. Twelfthly, that when there is nothing within the Pipe but Air, the whole weight of the Air L M becomes infenfible to me. The reafon is evident, becaufe it is wholly counterpoifed by the Air within the Pipe. I affirm thirteenthly, that the

the VVater E G, is in *equilibrio* with the Water A B ;
that's to say 1904 pound, is in *equilibrio* with 476 pound.
I prove it evidently, by the first *medium* ; all the parts of
an Horizontal surface, are equally prest ; therefore the
part A, sustains no more burden, then the part E, therefore A B, is as heavy as E G, and consequently, the Air
C D, must be as heavy, as the Air F H. Left this proposition may seem to contradict what is already said, I must
distinguish a twofold Ballance, according to the third Theorem, one *Natural*, another *Artificial*. In the *Artificial*
Ballance, where magnitudes do weigh according to all their
dimensions, *viz. Longitude, Latitude,* and *Profundity,* the
Water A B, and the Water E G, are not in *equilibrio* together, seing the one is 1428 pound heavier than the other.
But in the *Ballance* of *Nature*, such as these Pipes are, all
the four makes an *equipondium* together ; because they do
not weigh here, according to their *thickness*, but only according to their *altitude*. Therefore seing A B is as high
as E G, and seing C D is as high as F H, they must all
be of the same weight.

From the first assertion I infer, that one and the same
Fluid, even in the *Ballance* of *Nature*, may sometimes be
in *equilibrio* with a lesser weight, and sometimes with a
greater, because the Air C D, that weighs really 476
pound, is in *equilibrio* with the Water A B, that weighs
but 168. This is, when A B is supposed to be only 12
foot high. It's likewise in *equilibrio* with it, when its 34
foot high. But how can A B, that's 12 foot high, press
A, with as much weight, as when its 34 foot high? I answer by a similitude, when a Cylinder of Wood 12 foot
high stands upon a Table, it may burden it as much, as if
it were a Cylinder 34 foot high. For, supposing it to be
thrust

thruft in, between it, and *v. g.* the ceiling of the room
above, it muft prefs down with more weight, then if it
were not thruft in. So, this Cylinder of Water A B,
that's but 12 foot high, being preft between the furface A,
and the top of the Tub within, muft burden A, as much,
as if it were 34 foot high ; for being of this hight, it only
ftands upon the furface, without prefling up the top of
the Tub.

I infer from the fecond affertion, that each Pillar in a
Fluid hath a determinate weight. This is evident from
the determinate weight of A B, that weighs firft 168
pound, being 12 foot high, and 467 pound, being 34
foot high, and fo of the reft. I infer fecondly, that the
thicker, and groffer a Pillar of a Fluid be, it is the heavier,
(even in the *Artificial Ballance*) and contrariwife, the more
flender and thinner it be, it is the lighter. This is evident
from the Water A B, fix inches thick, that weighs 476
pound, and from the Water E G, 12 inches thick, that
weighs 1904 pound. So doth the Pillar of Air C D, weigh
lefs, then the Pillar F H. Here is ground for knowing
the certain and determinate weight of a Pillar, in any fort
of a Fluid whatfoever. As to Air, its clear and evident,
that a four-fquare Pillar thereof, 12 inches every way,
weighs 1904. That's to fay, if it were poffible, to take
the Pillar of Air F H, in its whole length, from the furface of the *earth*, to the top of the *Atmofphere*, and pour
it into the Scale of a Ballance, it would be exactly the
weight of 1904 pound. Here is a fecret : though that
fame Pillar of Air, were no longer , than 6 or 10 foot, yet
the Preffure of it, upon the body, it refts upon, is equivalent to 1904 pound. If this be, (you fay) what is the
weight of Air, that refts upon this Table, that's 36 inches
                                                    fquare ?

ſquare ? I anſwer, it muſt be as heavy, as a Pillar of Water 34 foot high, and 36 inches thick, which will, by juſt
reckoning, amount to 17136 pound, or to 1071 ſtone
weight. It may be inquired next, what's the weight of
the Air, that burdens the pavement of this parlour,
that's 16 foot ſquare ? I anſwer 487424 pound. Becauſe
it is exactly the weight of a bulk of Water 34 foot high,
and 16 foot thick. 'Tis to be remembred, that though
the Preſſure of it, be ſo much, yet being poured into
the ſcale of a Ballance, it will not weigh ſo much : for
not only as much as fills the room muſt be taken, but as
much as paſſeth from the pavement to the top of the
Atmoſphere. According to this method 'tis eaſie to determine the weight of any Pillar of Air whatſoever, provided
a man but once know the thickneſs of it, both the wayes,
e.g. there's a *planum* 12 inches long, and ſix inches broad,
upon which reſts a Pillar of Air. The weight of it then
is, juſt the burden of a magnitude of Water 34 foot in
hight, 12 inches in length, and ſix inches in breadth.

Though the weight of any Pillar of Air may be known,
by knowing only the dimenſions of it, in breadth and
length; yet the weight of a Pillar of Water cannot be
known, unleſs all the three common dimenſions of it,
be firſt known. The reaſon is this, the Pillars of Air,
are all of the ſame hight, but the Pillars of Water in the
Ocean, are of different hights : therefore, not only muſt
they be known, *ſecundum longitudinem*, *& latitudinem*,
in length and breadth, but *ſecundum profunditatem*, that
is, according to deepneſs. 'Tis eaſie to know then, what
each particular Pillar weighs. Firſt then, try how much
weight is in a cubical foot of Water, and having found
this to be *v.g.* 56 pound, you may determine, that a

Pillar

Pillar of Water 34 foot high, and 12 inches thick, weighs 1904 pound. A Pillar 34 foot high, and six inches thick weighs 476 pound. Note, that in a Cube of Water six inches thick, there are 216 inches, which weighs seven pound. In a Pillar 12 inches thick, and 20 fathom, or 100 foot high, you will find 5600 pound weight. In one, of the same thickness, but 200 fathom high, there are 56000, fifty six thousand pound weight. In a Pillar three foot square, and 20 fathom deep, there are 50400, fifty thousand, and four hundred pound weight. Make it 200 fathom high with that thickness, and it will weigh 504000, five hundred and four thousand pound. But, if according to the Theorem 25, you consider the weight of the Air above, it will weigh 521136, five hundred, twenty and one thousand, one hundred thirty and six pound. A Pillar 12 foot square, and 300 fathom deep, weighs 12096000, twelve million, ninety and six thousand pound, Lastly suppose there were a bulk of Water 500 fathom deep, and 500 fathom thick, such a magnitude would weigh 8750000000. eight thousand seven hundred, and fifty million of pounds. But if the Pressure of the Air, that rests upon a surface of Water 500 fathom in breadth and length, be taken in, that weighs 119000000, a hundred and nineteen million of pounds, the total, that the bottom of the sea sustains, must be 8940000000, eight thousand, nine hundred and fourty million of pounds, or 558750000 five hundred fifty and eight million, seven hundred, and fifty thousand stone weight.

I infer from the fifth assertion, that the lightest of Fluids may be brought to an *equilibrium* with the heaviest. For though Mercury be 14000 times heavier than Air,

yet

yet the part of the ſurface A, is no more preſt with
the Mercury A B, then the part C is preſt with the Air
C D. Secondly, that 29 inches of Mercury, are of the
ſame weight with 34 foot of Water. Thirdly, the heavier
a Fluid be naturally, it hath the leſs altitude in the
Natural Ballance; and contrariwiſe, the lighter it be, it
hath the more altitude. This is clear from the Mercury,
that's 29 inches, the Water that's 34 foot, and the Air,
that's counted 6867 fathom.

I infer from the ſixth aſſertion, that two Fluids of
different *gravities*, may make an *equilibrium* with a third
of the ſame kind. Becauſe the 27 foot of Air I G, and
the 17 foot of Water E K, are in *equilibrio* with the Air
F H. I infer ſecondly, that 17 foot of Air, may be as
heavy as 17 foot of Water, becauſe the Air I G, is exactly
as heavy, as the Water E K. I infer thirdly, that the
Benſil of a Fluid, is a thing really diſtinct, from the
Natural weight of it: becauſe the Preſſure of the Air I G,
is 952 pound; but the natural weight of it will not exceed,
if it were weighed in a Ballance, two or three ounces.
I infer fourthly, that Air cannot ſuffer dilatation, but it
muſt loſe of it's Preſſure. Becauſe the Air I G, that ought
to weigh 1904 pound, weighs only 952. For underſtanding this, you muſt know, that when a Pipe is
about half full of Air, and half full of Water, and inverted, ſo much of the Water falls out, and conſequently
ſo many inches doth the Air above it, expand it ſelf. So
to make this Pipe that's 34 foot high, half full of Air
and half full of Water, you muſt pour in about 19 foot
of Water, and the 15 foot of Air that's in it beſides,
will, when the Pipe is inverted, go up and expand it ſelf
to 17 foot, two foot of Water falling out.

I infer from the feventh affertion ; that when there are two Fluids of different gravities ; and weights counterpoifing a third, by what proportion the one grows lighter, by that fame proportion the other becomes heavier. For, when the VVater E K, that weighs 952 pound, becomes E N, that weighs 476, the Air above it, that weighed 952, becomes now 1428 pound.

I infer from the eighth, that the *pondus* of a Fluid, cannot be counterpoifed, by two diftinct *powers*. Becaufe the 34 foot of Water E G, cannot be both fuftained , by the part of the furface of Air E, and my hand. I infer from the ninth, that the Preffure and weight of a Fluid, may be found, even in its own Element, by *fenfe*. Becaufe in poifing of the Tub, I find the weight of the Air L M. I infer fecondly, that the *weight* of a Fluid is only found in its own Element, when there is not a *potentia* to counterpoife the *pondus* of it, becaufe I find only the weight of the Air L M, becaufe it wants a *potentia* to counterpoife it. I infer thirdly, that it is very poffible even in the *Artificial Ballance*, to weigh a Fluid in its own Element, and to know the precife weight of it, to a grain. For this caufe, take a fmall chord, and faften therewith the top of the Pipe G, to the Scale of a Ballance, and the Lead or Stone that makes the counterpoife in the oppofite Scale, is the juft weight of the Air L M.

I infer from the tenth, that by how much the nearer, the *potentia* of a Fluid, comes to the *pondus*, by fo much the lefs, is the *pondus* found, or is fenfible. This is clear, becaufe I find lefs of the weight of the Air L M, it being counterpoifed with the Air I G, than before. This follows likewife from the eleventh affertion. I infer from the twelfth, that when the *pondus* of a Fluid, is counterpoifed,

by

by an equal porosity, it becomes altogether insensible. I infer from the last, that two Fluids differing in weight, according to the *Libra* or *Artificial Balance*, may agree in weight according to the *Natural Balance*, I infer secondly, that Fluids in the *Balance of Nature*, do not counterpoise one another according to their thickness, but only according to their *altitude*.

To put a close to this Experiment, let us suppose the Pipe E G to be 68 foot high, and void of Air. If then the orifice E be dunned among stagnant Water, the Liquor of its own [first were] will rise from E to K 34 foot, the other half I G remaining empty. This evidently shews, that the Pressure of the Air, hath a *Sphere of Activity*, beyond which it is not able to raise or push up a pillar of VVater. 'Tis folly then to think that Water may be conveyed over high places by the help of a *Siphon*, v. g. from the one side of a Hill over the top, to the other side. For, if that height exceed perpendicularly 34 foot, no Art will do it. Yet considering, it is possible to transport Water, by *Pipes* and *Siphons*, not only 34 foot below the *source*, but 3400, Nay, if there were a *Siphon* passing from the surface of the Earth to the Center, and thence rising to the surface again, it would convey Water from the one place to the other. For 'tis a certain and infallible rule in the *Hydrostaticks*, that Water will rise as high as that place, to the height of the place it, from whence it comes, even though the windings and turnings of the *Lead-Conduits* underground were a *Labyrinth*, and though this place, were not only 3000, but 5000 foot distant from the other. 'Tis to be observed, that if the mouth of the Conduit

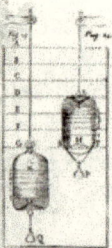

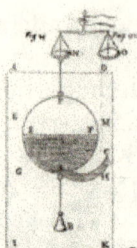

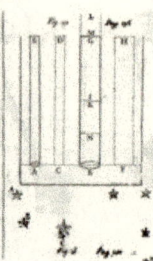

Bb 2   hee,

here, be exactly as high as the other end at the Fountain, the Water stands still. And the greater the difference be, the Water flows out with the greater force. By the help of such *Conduits*, 'tis easie to convey Water to a City many miles. Such Pipes are ordinarily made of Lead. But for saving expence, they may be made of Timber, or Clay well burnt in an Oven.

AN

AN ACCOMPT OF
Miscellany
OBSERVATIONS,
Lately made, by the Author of the foregoing Experiments.

OBSERVATION I.

IN *May* 1669, there was need of a new Sink, on the east side of *Tranent*, for winning of *Coals*. But while the *Coal-hewers* were in digging down, and had come the deepness of 13 or 14 fathom, they were stopped from working by *Damps*, or ill Air, that flowed out plentifully from the sides of the sink, wherein there were a great number of *Cutters*, or rifts, out of which that ill Air came. To try the nature and power of *Damps*, I took a dog, and fastned him in a *bucket*, with a small roap, that he might not leap over, and when he had gone down 7 or 8 fathom, he presently begins to howl, and cry pitifully, as if he had been

beaten

beaten ſore with a rod, and a little after, he begins to ſtag-
ger, and his feet failing him, he falls down, as one overtaken
with the Epilepſy, and in going down to the bottom, his
eyes turning in his head, they appeared very ſhining and
clear like two large bright Diamonds. Fearing, that the
Damp ſhould have killed him out of hand, he was inſtantly
pulled up from the bottom, where he had not tarried 15 ſe-
conds of time. And when the bucket had come to the mouth
of the *ſink*, he was pulled out, and laid upon the ground, to
get freſh Air. When he had lien a while as dead, he begins
at laſt to gape, and gaſp, and make ſome reſpirations, as if he
had been rather expiring, than recovering. Next, he began to
ſtir and move his feet, and after, to raiſe himſelf upon his
knees, his head ſtaggering and wavering from ſide to ſide.
After a *minut* or two, he was able to ſtand upon his feet,
but ſo weakly, that he was not in capacity to walk or run.
Yet at laſt, being much refreſhed, he eſcaped from us, and ran
home, but ſlowly. In the afternoon, the ſame Experiment
was repeated, with another dog, whoſe caſe was the ſame
in all things. But after he was perfectly recovered, for a
further trial, we let him down the ſecond time, and ſuffered him to tarry in the bottom of the *ſink*, about the ſpace
of three minuts: but when he was pulled up, and taken
out, we found no ſymptomes of life in him; and ſo after
half an hour and more, his body began to ſwell, which ordinarily befalls ſuch, who are killed after this manner. After this, we ſent down in the Bucket, a little Chicken,
which, when it came near the *Damp*, preſently flapped
with the wings, and falling down, turned over and over
for a pretty while, as if it had been taken with a *vertigo*, or
giddineſs. But by drawing up the Bucket in haſte, and
bringing the Bird to the freſh Air, it recovered. In the
evening.

evening, we let down a *lighted Candle*, but it was soon
extinguished, when it came near mid-sink; for here, rather than in the bottom, was the strongest Damp. Lastly, we let down by a chord, a *Brand-iron*, with burning
Coals, whose flame was soon put out, and after a little
while, we perceived the red Coals to be extinguished by
degrees; yet not totally, because, as the Coal-hewers observed, the power of the *Damp* was not so strong, as before.
These *Damps* then have their ebbings and flowings, which
seem to depend upon the weather, or rather upon the situation of the winds, and their force. For 'tis observed, that a
high South-west wind causeth ill Air in this place; and
that, by reason of much waft ground, that lies upon the
South, and South-west hand of this *Sink*, whence are conveyed under ground by secret passages, which are nothing
else but so many rifts and openings, commonly called by
the *Coal-hewers*, *Cutters*, corrupted and rotten Air, full
of sulphurious stems. The reason why these passages are
open, and replenished with nothing, but corrupted Air, is
this, the Water, that's ordinarily called the Blood of the
Coal, being withdrawn with subterraneous Gutters (commonly called *Levels)* that are digged, and wrought under
ground, sometimes a very long way, for drying of the *Mines*,
and the veins of the earth being now empty, there succeeds Air; which Air, by process of time, and long standing, rots, and contracts a sulphurious quality, which
causeth sudden death. Now, when the wind is high, and
strong from the South or South-west, that sulphurious Air
is driven through the ground, and coming to *Sinks* and
Mines, where men are working, presently infects the place,
and hinders the work. 'Tis often observed, that the wind
and Air under ground, keep a correspondence in their moti-

on,

on, with the wind above ground : and therefore, when the
wind is in such a point above, 'tis found, that the motion
of the Air below runs such a way, and the contrary way,
when the wind above ground, is in the opposite point.
When there is a free passage between the bottom of the
two Sinks, you may observe the wind come down through
the one, and running alongst under the ground, rise up thorow the other, even as Water runs thorow a *Siphon*. For
this cause, when the *Coal-hewers* have done with such a
Sink, they do not use to stop it, or close it up, but leaves
it standing open, that the Air under ground may be kept
under a perpetual motion and stirring, which to them is a
great advantage. 'Tis very strange to see sometimes, how
much Air, and how fresh it will be, even at a very great
distance, namely four or five hundred pace, from the mouth
of the *Sink*. This could never be, unless there were a considerable Pressure and weight in it, whereby it is driven forward, thorow so many *Labyrinths*. And even in the utmost
room, where the *Coal-hewers* are working, the Pressure is as
great, as it is above ground, which is found by the *Torricel-*
lian Experiment. In such a case, the Air cannot press down
thorow the Earth and Metalls, therefore the Mercury must
be suspended, not by a Pillar from the *Atmosphere*, but by
the *Bensil* of it. Nay, put the case, that the whole Element
of Air were destroyed, and this remaining, yet would it be
able to support 29 inches. To shut up this discourse, it is observed by the *Coal-hewers*, that when there is ill Air in a
Sink, a man may perceive distinctly, what is lying in the bottom, so clear and transparent is the Air of it: but when the
Damp is gone, the *Medium* is not so clear. In temperat and
cold weather, the *Damps* are not so frequent. From this *Sink*,
in soft winds, or in Northerly winds, or when it blows from
East or North-east, the *Damps* are driven away.   O B-

OBSERVATION II.

Jupiter upon Wednesday night , at eleven a clock, being 24 of *November* , 1669 , had the following position with the stars of *Gemini*. He was so near to the Star C, that to appearance , the points of his rayes did touch it. This Star by looking upon the material Glob, is fixed in the very *Zodiack* , and in the 13 degree of *Cancer*, and is the very navel of the following *Twine*. The Star A is *Castor*. The Star B is *Pollux*. The star D , is fixed in the forefoot of the following *Twine*. From this place he moved , with a retrograde motion , till he came to the 5 of *Cancer* , about the 20 of *February* , 1670, and from that time became *Direct* in his motion , and so upon the 27 of *March* , 1670 at 9 a clock , he was in a right line with *Canis minor*, and the brightest Star in *Auriga*, and was in a right line with the eastmost shoulder of *Orion* , and *Castor* in *Gemini* , or with that Star , when South-west, that's highest, and West-most.

OBSERVATION III.

IT is written in the History of the *Royal Society* , that such a member of it , whose name I have forgotten, hath found out , among many other curious inventions, this, namely a way for knowing the motion of the Sun in *seconds* of time : but is not pleased to reveal the manner how. Because such a *device* may be usefull in Astronomy, and likewise for adjusting the *Pendulum Clock* , I shall therefore briefly shew, the manner and way how such a thing may be done , as I have tried it my self. I took an *Optick Tub*, about 12 foot long , only with two *Convex-*

glasses

glasses in it, and did so place it in a dark room, by putting
the one end, in which was the *object-glass*, without the
window, and keeping the other within, that I caused the
beams of the Sun shine thorow it, which were received upon
a white wall four or five foot from the Tub. This image,
which was perfectly round, and splendid, did move
alongst the wall very quickly, so that in a minut of time,
it did advance seven inches and a half, which will be the
eight part of an inch in a *second*, a motion very sensible.
Now, this beam that came thorow the Tub, and lighted
upon the wall, would not have moved one inch in a minut,
if it had wanted the two Glasses; for as they magnify,
and seem to bring nearer the *object*, so they quicken the
motion of it. In a word, by what proportion the *object*
is made more, by that same proportion is the motion
quickned. 'Tis to be observed, that the longer the Tub
be, the motion is the swifter: for as the longest Tub
doth ordinarily most magnify the object; so doth it most
quicken the motion. Next, the farther distant the
white wall is from the end of the Tub, the larger is the
image; and contrariwise, the nearer it be, it is the less.
Thirdly, the farther the wall be from the end of the
Tub, the circumference of the image is the more confused, and the nearer it be, it is the more distinct.
Fourthly, the darker the room be, it is so much the
better. Lastly, this trial may be made with ordinary
Prospects, of a foot, two foot, or three foot long, which
will really do the thing, but not so sensibly, unless the
glasses be very good.

As to the use of this device in Astronomy; I shall not
say much. But shall only mention what it may serve for
in order to the Pendulum Clock. For this cause, let a

man choise a convenient room, with a window to the South, wherein this Tub may be so fixed, that it may ly just, or very near to the true *meridian*, and may move *vertically* upon an axil-tree, because of the Suns declination every day. Then at a certain distance from the end of it, fix and settle a large board of timber, smooth, and well plained, and well whited, for receiving the image. In the middle of this board, draw a circle with Charcoal, equal in diameter to the circle of the image. Now, this being done, you will find that assoon as the west side of the Sun, begins to come near to the *Meridian*, the image begins to appear upon the board, like the segment of a circle, and grows larger, and larger, till it become perfectly round. Now in the very instant of time, wherein the image, and the circle are united, set the wheels of your Clock a going, from the *hour*, *minut*, and *second* of XII. To morrow, or 3 or 4 dayes after, when you desire to make an examination, wait on about 12 a clock, when the Sun is coming to the *Meridian*, and you will find what the difference is. If the *Clock* go slow, observe, assoon as the image is united with the circle (which you will perceive in a *second* of time) the variation, that's to say, how many *second*s interveens between that *second*, wherein the union fell, and that *second*, that closes XII hours in the *Clock*. If it go fast, observe how many *seconds* passes from that *second*, that ends XII hours, and that wherein the image of the Sun is united with the circle, which if you do, you will know exactly, what the difference is, even to a *second*. But without this, you will find great difficulty to know the variation in 15 or 20 *seconds*, especially in a common *Dial*. But here, you will see distinctly the image of the Sun move every *second*

of

of time, the eighth part, or the sixth part, or the fourth
part of an inch, according to the length of your Tub, and
goodnefs of your glaffes. 'Tis to be obferved, that in
adjufting the *Pendulum Clock*, refpect muft be had to
the table of *Equation* of *dayes*, commonly known in
Aftronomy. For if this be not, it is impoffible to make
it go right, and that becaufe all the natural dayes of the
year, are not equal among themfelves: that's to fay, the
time that's fpent by the Suns motion from the *Meridian*
this day, to the fame *Meridian*, the next day, is not
equal, but is more or lefs, than the time fpent betwixt
Meridian and *Meridian*, a third or fourth day after. For
inftance, the Sun this day being 11 of *July*, comes fooner
to the Meridian by three *feconds* of time, than he came
yefterday. Within 9 or 10 dayes, (fuppofe the 22 of *July*)
he will be longer in coming to the Meridian by 4 *feconds*,
than upon the 21. This difference I grant, in fhort time
is not fenfible, yet once in the year, it will amount to
more than half an hour. This inequality of dayes arifes
from two caufes. Firft, from the Suns *eccentricity*, whereby
he moves flowlier in one part of the Zodiack, than in
another: for in Summer when he is furtheft from the
Earth, he goes flowlier back in the Ecliptick, than in
Winter, when he is nearer to it. The fecond caufe,
which is truly the far greater, is this, becaufe in the
diurnal motion of the Sun, equal parts of the *Æquator*,
does not anfwer to equal parts of the *Zodiack*. Hence
it followes, that if the natural dayes be not equal among
themfelves, the hours muft be unequal alfo: but this is
not confiderable.

By help of fuch a Tub placed in a dark room, it is
eafie, when the Sun is under Eclipfe, to enumerat diftinctly

ly the digits eclipſed. Likewiſe, if you take out the
objeƈt Glaſs, and cover a hole in the window board with
it, you ſhall ſee diſtinƈtly upon a white wall, the *ſpecies*
and true repreſentations of all objeƈts without. And by
comparing the quantity of the objeƈt without, with the
quantity of it within, you may know the diſtance of it
from the window; though it were many miles. For as the
one quantity, is to the other, ſo is the diſtance between
the Glaſs and the objeƈt on the wall, to the diſtance between the Glaſs and the objeƈt without.

It may be inquired whether or not, the *retrograde*, as
well as the *diurnal motion* of any of the Planets, may
be diſcerned, in *minuts* or *ſeconds*, by the help of a long
Teleſcope ? In anſwer to this, we muſt ſuppoſe the
Planets only to have a *retrograde* motion, and conſequently
to move ſlowly from Weſt to Eaſt; *Saturn* once in 29
years, or 30, to run about the *Zodiack*; *Jupiter* in 12,
Mars in 2 years, the Sun in one year, *Venus* and *Mercury*
in leſs time, and laſtly the *Moon* in a moneth. Now I
ſay, it is impoſſible by the longeſt Tub, that the greateſt
Artiſt can make; to diſcern the motion of the inferior
Planets, far leſs the motion of the ſuperior, either in
Minuts or in *Seconds*, and that by reaſon of the great
tardity, and ſlowneſs of the motion. Notwithſtanding
of this, I am induced to think, that the retrograde motion
of the *Moon* might be diſcerned, at leaſt in *Minuts*. For
evincing of this, let us ſuppoſe which is true, that the
Sun runs from Eaſt to Weſt half a degree in two Minuts
of time, ſeing in an hour he runs 15 degrees. Next, that
the *Moon* goes about the *Zodiack* in 27 dayes and 7 hours,
namely from that ſame point, to that point again, and
conſequently runs back every day 13 degrees and about

10

10 Minuts. By this account, she muſt *retrograde* half a degree, and about 2 minuts of a degree every hour. The *Sun* then runs half a degree in two *Minuts*, and the *Moon* half a degree in 60 *Minuts*; therefore the *Moon* muſt be 30 times ſlower in her *retrograde* motion, than the *Sun* is in his diurnal motion. Let us ſuppoſe next, as I obſerved with a Tub 12 foot long, that the image of the *Sun* runs the eighth part of an inch every *ſecond*, and conſequently, ſeven inches and an half, in a *Minut*: then muſt the image of the Moon with that ſame *Teleſcope*, run the thirtieth part of ſeven inches and a half in a Minut, ſeing ſhe runs 30 times ſlowlier; therefore in every *Minut* of time ſhe muſt advance the fourth part of an inch, which will be very ſenſible. Though we grant, that the *Moon* hath no *retrograde* motion properly, yet by comparing the diurnal Motion of the *Moon*, that's ſlower, to the diurnal motion of the *Sun*, that's ſwifter, we ſhall really find the thing it ſelf. Therefore in the time of a *Solar Eclipſe*, this *retrograde* motion is conſpicuous, which by an ordinary *Teleſcope* may be diſcerned in *Minuts*. Aſſoon then as the Eaſt ſide of the *Moon*, begins to enter upon the Weſt ſide of the *Sun* (the greater the *Eclipſe* be, it is the better) obſerve, and you will find the one image, which will be black, cover the other by degrees, that's ſplendid, and run in every minut of time, the fourth part of an inch of the *Suns* diameter, provided alwayes, that the *Sun* run the eighth part of an inch in a *ſecond*.

O B S E R-

OBSERVATION IV:

UPon *Tuesday* the 19. of *July* 1670, the following Experiment was made. In the middle Marches between *Scotland* and *England*, there is a long tract of Hills, that run from *Flowdon*, many miles South and South-west, amongst the which, the Mountain *Cheviot* is famous beyond, and conspicuous above all the rest for altitude, from whose top a man may discern with one turning of his eye, the whole Sea-coast from *New-castle* to *Berwick*, much of *Northumberland*, and very many Leagues into the great *German Ocean*: the whole *Mers* and *Teviotdale*, from the foot of *Tweed*, to very near the head of it: *Lauderdale*, and *Lammer-moor*, and *Pentland hills* above *Edinburgh*. The North side of this Mountain is pretty steep, yet easie to climb, either with men or horse. The top is spacious, large and broad, and all covered with a *Flow-moss*, which runs very many miles South. When a man rides over it, it rises and falls. 'Tis easie to thrust a Lance over the head in it. The sides of this Hill abounds with excellent Wellsprings, which are the original of several Torrents, amongst the which *Colledge-Water* is famous, upon which, not a mile from the foot of this Mountain is *White-hall*. The adjacent Hills are for the most part green, and excellent for the pasturage of Cattel. Not many years ago, the whole Valleys near the foot of *Cheviot*, were Forrests abounding with *Wild-Deer*.

Upon the highest part of this Mountain was erected the *Torricellian Experiment* for weighing of the Air, where we found the altitude of the *Mercurial Cylinder* 27 inches and an half. The Air was dry and clear, and no wind. In our Valley-Countreys, near to the Sea-Coast, in such

Weather,

Weather, we find the altitude 29 inches and an half.
When this difference was found, care was taken to seal up
clofly with *Bee-wax*, mixed with *Turpentine*, the orifice
of the Veffel, that contained the ftagnant Mercury, and
thorow which the end of the Pipe went down. This being
done with as great exactnefs as could be, it was carried to
the foot of the Mountain in a Frame of Wood, made on
purpofe, and there opening the mouth of the Veffel, we
found the Mercury to rife an inch and a quarter higher than
it was, The reafon of this ftrange *Phenomenon* muft be this,
namely a greater Preffure of the Air at the foot of the Hill,
than upon the top: even as there is a greater Preffure of
Water in a furface 40 fathom deep, than in a furface 20
fathom deep. 'Tis not to be doubted, but if the root of
the Mountain had been as low as the Sea Coaft, or as the
furface of *Tweed* at *Kelfo*, the *Mercurial Cylinder* would
have been higher. This way of obferving, feems to be
better than the common: for while the *Barofcope* is carried up and down the Hill, without ftopping the orifice of
the Veffel, that contains the ftagnant Mercury, the *Cylinder* makes fuch reciprocations, by the agitation of a mans
body, that fometimes abundance of Air is feen to afcend
up thorow the Pipe, which in effect makes the *Cylinder*
fhorter than it ought to be. But if fo be, the end of the
Pipe be immerged among *Quick-filver*, contained in a
Glafs with a narrow orifice, fo that it may be ftopped compleatly, you will find no reciprocations at all. And to
make all things the more fure, the Glafs may be filled up
either with *Mercury*, or with Water above the *Mercury*;
by which means the *Cylinder* in the down-coming, or in the
up-going fhall remain immoveable. Befides the ftopping
of the orifice of the faid Glafs, you may have a wider Veffel,

sel, that may receive the same Glass into it, and it being full of Water, may so cover the sealed orifice, that there shall be no hazard of any Air coming in. Or this Experiment may be first tried at the root of the Hill, and having stopped compleatly the mouth of the Vessel, the whole Engine may be carried up to the top, where you will find the *Mercury* subside and fall down so much ; namely after the said orifice is opened : for as the stopping of the orifice at the root of the Hill, is the cause, why that same degree of Pressure remains in the stagnant Liquor ; so the opening of it upon the top of the Hill, is the cause why it becomes less.

This Experiment lets us see, that the Pressure of the Air seems to be as the Pressure of the Water, namely the further down the greater ; and the further up the less : and therefore, as by coming up to the top of the Water, there is no more Pressure, so by coming up to the top of the Air, there is no more weight in it ; which in effect sayes, that the Air hath a determinat hight, as the Water hath. From this Experiment we cannot learn the determinat hight of the Air, because the definit hight of the Mountain is not known. I know there are some, who think that the Air is indefinitly extended, as if forsooth, the Firmament of fixed Stars were the limits of it, but I suppose it is hard to make it out.

OBSERVATION V.

June 5. 1670. I observed the *Sun* within 3 *minuts* of setting, to have a perfect *oval figure*, the two ends lying level with the Horizon. His colour was not red as ordinarily, but bright and clear, as if he had been in the

D d

Meridian :

Meridian : neither was the Sky red, but clear alfo. And by the help of the *Pendulum Clock*, I have obferved his body to be longer in fetting than it ought, by eight *minuts*, and fometimes by *ten*, and his Diameter longer in going out of fight than it ought, by two, and fometimes by three *minuts*. The reafon of thefe *Phenomena*, muft be the *Refraction* unqueftionably.

OBSERVATION VI.

UPon *Saturday* evening the 30 of *July* 1670, and the night following, till about two a Clock in the *Sabbath* morning, there fell out a confiderable rain, with great thunder, and many lightnings. About *Sun-fet*, the convocation of black clouds appeared firft towards the *Horizon* in the South-weft, with feveral lightnings; and the wind blowing from that point, carried the clouds and rain over *Mid* and *Eaft-Lothian*, towards the *Firth* and *Seacoaft*. About 9 a clock, the whole Heavens almoft were covered with dark clouds, yet the rain was not very great, neither were the *thunder claps* frequent, but every *fifth* or *fixth fecond* of time, a large and great lightning brake out. But before the *thunder crack* was heard, which happened every fourth or fifth *minut*, the lightning was fo terrible for greatnefs, and brightnefs, that it might have bred aftonifhment. And becaufe the night was very dark, and the lightning very fplendid, a man might have perceived houfes and corn-fields at a great diftance. And if any had refolved to catch it, in the breaking out, it did fo dazle the eyes, that for half a *minut*, he was not able to fee any thing about him.

Sometimes the lightning that went before the thunder, brake

brake forth from the clouds, like a long spout of fire, or rather like a long flame raised high, with a Smiths Bellows, but did not continue long in sight. Such an one above the Firth was seen to spout downward upon the Sea. Sometimes there appeared from the one end of the cloud to the other, an *hiatus*, or wide opening, all full of fire, in form of a long furrow, or branch of a River, not straight, but crooked. I suppose the breadth of it, in it self, would have been twenty pace and more, and the length of it five or six hundred pace : the duration of it, would have been about a second of time. Sometimes a man might have perceived the nether side of the cloud, before the crack came, all speckled with streams of fire, here and there, like the side of an Hill, where Moor-burn is, which brake forth into a lightning. But there was one, after which followed a terrible thunder crack, which far exceeded all the rest, for quantity and splendor. It brake out from the cloud, being shot from North to South, in form of fire from a great *Cannon*, but in so great quantity, as if a Gun ten foot wide, with 500 pound weight of Powder in it, had been fired. And surely the lightning behoved to be far greater in it self, seeing it appeared so great, at so great a distance. It did not evanish in an instant, like the fire of a Gun, but continued about a second and an half ; by reason (it seems) that it could not break out all at once. This did so dazle the sight, that for half a minut almost, nothing was seen, but like a white mist flying before the eyes. The whole Countrey about was seen distinctly.

All these great lightnings were seen a considerable time, before the crack was heard. Sometimes 30 *seconds* numbered by the *Pendulum Clock* interveened , namely when the thunder was at a distance, about 7 or 8 miles. Some-

times

times 15 or 16 only interveened. But when the thunder was juſt above our head, no moe paſſed, than 7 or 8, which ſeems to demonſtrat, that theſe thick black clouds, out of which the thunder breaks, are not a *Scottiſh* mile from the earth, when they are directly above us.

'Tis obſervable, that in all lightnings, and thunderings, there is no ſmoke to be ſeen, which ſeems to evince, that the matter whereof they are generated, muſt be moſt pure, and ſubtil. Who knows, but this Countrey, that abounds with *Coal*, may occaſion more thunder and lightnings, than other places, namely by ſending up ſulphurious exhalations to the middle region of the Air, wherewith the *Coal-mines* abound.

OBSERVATION VII.

THis is a method for finding out the true South and North Points, which are in effect very difficult to know. Take therefore four pieces of Timber, each one of them five foot long, and about ſix inches thick, ſquare-wiſe. Sharpen their ends, and fix them ſo in the ground, that they may ſtand Perpendicular, and as near to South and North, by a *Magnetick Needle*, as may be. The place would be free of Trees, or of any ſuch impediment, that it may have a free proſpect of the Heavens. As for their diſtance one from another, let the two North-moſt, and the South-moſt be two foot aſunder: let the two Eaſt-moſt, and two Weſt-moſt, be but one foot, making as they ſtand, an *oblong quadrangle*. For keeping them equi-diſtant above, as well as below, take four bars of Wood, about three inches broad, and one inch thick, and nail them round about upon the four ſides, on each ſide one, ſo that

being

being nailed on *Horizontally*, they may make *right angles*, with the tops of the standards above. There are then for distinctions cause, the North-bar, and the South-bar, that runs East and West, and the East-bar, and the West-bar, that runs South and North, There is here no difficulty in the thing it self, but only in the fancy to conceive it. Besides these four, there must be other four of the same form and fashion, nailed on 'arder down about the middle of the four standards. Take next some small Bra's Wyre strings, such as are used in *Virginals*, and fix one from the middle of the South-bar, that's upmost, to the middle of the South-bar just under it. Fix it so, that it may be exactly Perpendicular, which may be done, with a great weight of Lead. Take a second Wyre string, and hang it plumb from the West end of the North-bar, and another from the East end of the same Bar, I mean the Bar that's nearest to the top. These three strings so fixed, will go near to make an *equilateral triangle*.

Now because the device is for finding out the *Meridian* by the Stars in the night time, not by any indifferently, but by these that are nearest to the *Pole*, therefore observe in *July* and *August*, when the *Guard-stars* in the evening begin to come down towards the West, and keeping clois one eye, bring the other somewhat near to the South-most string, and order your sight so, that this string, and the West-most string upon the North side, may catch the foremost *Guard-star* in the down-coming, when it is furthest West, and there fix it. When the same Star is turning up towards the East, catch it by the South-most string, and the East-most string on the North side, and your work is done, if so be, you divide exactly, between the East-most and West-most, and there hang a fourth string, which

with

with the string upon the South-side, gives you the true
South and North. For better understanding, note first,
that, when the *Guard-stars* are coming down, or going up,
the *Altitude* varies quickly, but the *Azimuth*, or motion
from East to West, will not vary sometimes sensibly in two
hours almost, which is a great advantage in this case. But
when you find out the *Meridian* with a *Plain*, and a Perpendicular *Stilus*, by the shadow of the Sun, if it be not when
he is about East and West, the *Azimuth* alters mo.e than
the *Altitude*, wh ch is a great disadvantage. Now its
certain, the slower the motion be from East to West of any
Star, it is the easier to observe, and it is the more sure
way. Note secondly, that special care must be had, to
cause the strings hang Perpendicular. Note thirdly, that
before you begin your Observations, the South-most string
must be made immoveable, but the East-most, and West-
most, on the other side, must not be so, because as the
Stars in going about move from East to West, so must the
said two strings be left at liberty, to move a little hither
and thither, till the Observations be ended. Note fourthly, that assoon as you perceive sensibly, the foremost *Guard-*
star to decline towards the West, then you must begin to
observe, which is nothing else, but to fix your eye so,
that the South-most and West-most string, may cover the
said Star. And because in coming down, it goes West,
therefore, let the West-most string move towards the left
hand by degrees ; following the Star to its utmost, till it
be covered by them both. Follow the same method, in
observing the same Star in going up towards the East. Note
fifthly, that when you make the two strings cover the Star,
that which is nearest to the eye, will appear transparent,
and of a larger size, so that you may perceive distinctly
thorow

thorow it, not only the Star it self, but the other string also, which is a great advantage. This is evident to any,
who holds a bended silk threed between their eye and a
Star in the night time; for when you direct your sight to
the Star, the string appears like the small string of a *Vir-*
ginal when it trembles. Note sixthly, that in observing
in a dark night, you must have a *Cut-throat,* that by the
light of the candle you may perceive the strings. Some
other things might be noted, but you will find them better by experience, than they can be exprest here.

I named *July* and *August* in the evening for observing
the *Guard-stars,* when they are West-most, but there are
several other seasons, when this may be done as conveniently. They are East-most in the latter end of *October,* and
beginning of *November* about 5 or 6 a clock in the morning.
If a man were desirous to make this observation quickly, I
suppose he might in the end of *October,* find the said stars
West-most in the evening, and East-most the next morning.
Besides the *Guard-stars,* a man may make use of the *Polar-*
star; for as it goes higher, and lower than the true *Pole,* by
2 degrees and 26 minuts, so it goes as much to the East,
and as much to the West, once in 24 hours. In the end
of *July,* you will find the *Polar-star* East-most, about 9
a clock at night, and in the end of *January* West-most at 9
a clock. Note, that every month, the fixed stars come
sooner to the same place by two hours: therefore in the
end of *August* the *Polar-star* must be West, at 7 a clock at
night, and East at 7 a clock in the morning. When the
Meridian is found out after this manner, there is no *Star* or
Planet can pass it, but you may know exactly when, be it
never so high, or never so low. For there is nothing to
be done, but to wait, till the South-most and North-
most

moſt ſtring cover the body of the *Star.* If it be the Sun, hold up a white Paper, behind the two ſtrings , and when their ſhadows do co-incide, and are united , then is his Center in the *Meridian.* If the Sun do not ſhine clear, as when he is under miſt, or a thin cloud, you may exactly take him up in the *Meridian*, with the two ſtrings. This Frame will ſerve as well , to know when any of the North Stars comes South, or North, and conſequently when they are higheſt, and when they are loweſt : for being fixed in an open place of the *Orchard,* there's no *Celeſtial Body* can paſs the *Meridian,* either on the one ſide, or the other, but it may be catched, what ever the Altitude be, and that moſt eaſily.

OBSERVATION VIII.

THere hath been much inquiry made by ſome anent the reaſon, why the dead body of a man or beaſt, riſeth from the ground of a Water, after it hath been there three or four days. But though many have endeavoured to ſolve the queſtion, yet the difficulty remains ; and in effect it cannot be anſwered, without the knowledge of the fore-going Doctrine, anent the nature of fluid Bodies. To find out the reaſon then of this *Phenomenon,* conſider, that all Bodies, are either naturally heavier then Water, as Stone and Lead, or naturally lighter, as Wood and Tim-ber. If they be heavier, they ſink : if they be lighter, they ſwim. Now I ſay, a mans body immediatly after he is drowned, his belly being full of Water, muſt go to the ground, becauſe in this caſe, it will be found *ſpecifi-cally* or *naturally* heavier then Water. That's to ſay, a mans body, will be heavier, than as much Water, as is the

the bulk of a mans body. For pleasing the fancy, imagine a Statue to be composed of Water, with all the true dimensions of the person that's dead, so that the one shall answer most exactly to all the dimensions of the other. In this case, if you counterpoise them in a Ballance, the real body, that's made up of flesh, blood, and bones, shall weigh down the other. But after this dead body hath lien a short time among the Water, it presently begins to swell, which is caused by the fermentation of the humors of the blood, which goeth before putrefaction, and after three or four dayes swells so great, that in effect, it becomes naturally lighter than Water, and therefore riseth. That is to say, take that body, that is now swelled, and as much bulk of Water, as will be the precise quantity of it, and having counterpoised them in a Ballance, you will find the Water heavier than the body.

OBSERVATION IX.

UPon *Thursday* the 25 of *August* 1670, the following Experiment was made in a new *Coal-sink*, on the West side of *Tranent*. When the *Coal-hewers* had digged down about 6 or 7 fathom, they were interrupted sometimes with *ill Air*: therefore to know the power and force of the *Damp*, we let down within the Bucket a *Dog*. When he had gone down about 4 fathom, or middle Sink, we found little or no alteration in him, save only that he opened his mouth, and had some difficulty in breathing, which we perceived evidently: for no sooner he was pulled up to the top, where the *good Air* was, but he left off his gaping. We let him down next to the bottom, where he tarried a pretty while, but no more change we found in

E e   him

him than before. After this we let down a great qu.ntity of *Whins*, well kindled with a bold flame, but they no fooner came to the middle of the *fink*, but the flame was in an inftant extinguifhed: and no fooner was the Bucket pulled up, but they took fire again. This was 5 or 6 times tried, with the fame fuccefs. If we compare this Obfervation with the firft, we will find, that all *Damps* are not of the fame power and force; but that fome are ftronger, and kills men and beafts in an inftant: and that others are lefs efficacious, and more feeble, and doth not fo much hurt, and that men may hazard to go down into a *Sink*, where *ill Air* is, even though fire be fometimes extinguifhed. We fee next, that thefe *Damps* doth not always infect the whole Air of a *Coal-pit*, but only a certain quantity: for fometimes it is found in the bottom, fometimes in the middle. And we fee laftly, that they are not alwayes of long continuance: for it is found, that though the Air be ill in the morning, yet it may be good ere night; and totally evanifhed ere the next day. We may add, as was noted in the firft Obfervation, that thefe *Damps* depend much upon the fcituation of the winds, feing in ftrong Southerly winds, they are frequently in thefe places.

OBSERVATION X.

OF thefe many excellent devices, that have been found out of late, the *Air-pump* is one, firft invented in *Germany*, and afterwards much perfected in *England* by that Honourable Perfon Mr *Boyl*, who for his pains, and induftry in making Experiments therewith, deferves the thanks of all learned perfons. Several trials hath been made of late by it, fome whereof, are as follows. I took a flender

Glafs-

,o inches long, cloſs above, and open be-
with VVater. I next inverted it, and
:, juſt upon the mouth of the Braſs-pipe,
d thorow the board, whereon the *Receiv*-
and cemented them together. At the
ie whole VVater in the Pipe fell down,
ie *Braſs-conduit* to the Pump. Having
topped the paſſage, and thruſt down the
ened it again, and the *Pump* being full of
riven with a conſiderable force up thorow
is it not compleatly fill'd as before, by
r, that I ſaw in the top. After this was
ire five or ſix times, I opened the *Stop*-
, than I had uſed, but the VVater, by
ſo furiouſly driven up thorow the Tub,
ʒroke the end of it, that was *Hermetically*
ʒiece that flew off, did hit the ſeiling ſo
ʒounded a very far way. From this we ſee
VVater falls not down from Veſſels that
:s, though they be inverted, becauſe it's
:e and power of the environing Air. 'Tis
though this Pipe had been 30 foot high,
ater in it would have ſubſided, and fallen
exſuction.
was with the help of a ſmall *Receiver*,
s a real *Cupping-glaſs*. This had a hole
ʒm of it, and was cemented to the *Braſs*-
ʒuth of it looking upward, had a lid for
I took next the lately mentioned Glaſs-
with good *Brandy*, and having drowned
ig ſtagnant *Brandy*, I ſet the Veſſel where-
he *Receiver*, the Pipe coming up thorow

Ee 2   the

the lid, and having cemented it closly, I made the first exsuction, and found no descent of the Liquor from the top of the Tub. At the second, it fell down about an inch. At the third, it fell down four or five. But here appeared a great multitude of small Bubbles of Air, like broken VVater, near the top of the Pipe within. And besides this *Phenomenon*, there ascended from the stagnant Liquor up thorow the Pipe, an infinit number of small Bubbles, no bigger than Pin-heads, for a very large time. VVith a fourth exsuction, it fell down within two or three inches of the *stagnant Brandy*. And thinking to make the one level with the other, I made a fifth; but here appeared a strang effect, namely, not only the whole *Brandy* in the Pipe subsided, and was mingled with the *stagnant Brandy*, but at this exsuction, there came a great quantity of Air from the mouth of the Pipe, and rose up thorow the *stagnant Liquor* in Bubbles. Having made another exsuction, there came yet more Air out, and so copiously, that I thought there had been some leak in the Tub, through which the outward Air had entered; but knowing the contrary, I continued Pumping a very long time, till I found less and less come out, and at length, after near 30 exsuctions it ceased. This Air to appearance, was so much as might have filled twenty Tubs, every one of them as large, as the Tub it came out of. And surely all of it came out from among the small quantity of *Brandy* that filled the Pipe, and that environed the mouth of it, I mean the *stagnant Brandy*, both which would not have been eight spoon- ful. After this I opened the *Stop-cock* leasurely to let in the Air to the *Receiver*; then did the *Brandy* climb up the Pipe slowly, till it came near to the top, and there made some little halt, by reason of half an inch of Air that ap-

peared

*peared there. But more and more Air coming into the *Receiver*, that half inch in the top of the Pipe, did so diminish, that it appeared no bigger than the point of a Pin, and was scarcely discernable to the eyes. What a strange and wonderful faculty of dilatation and contraction must be in the Air, seing that which presently had filled the whole Tub, that was 40 inches long, and the sixth part of an inch wide, was contracted to as little room, as the point of a Needle. And by making some new exsuctions, that small *Atome* of Air did so dilate it self again, that it filled the same Tub, and not only that, but, as formerly, it bubbled out from the mouth of the Pipe several times.

'Tis to be observed, that though at the first falling down of the *Brandy*, it appeared like broken Water, near the top of the Pipe within, yet no such thing was seen the second time it fell down; the reason is, because by the first exsuctions, it was well exhausted of its *aërial particles*. Once or twice I found, after the *Brandy* within the Pipe was well freed of Air, that no exsuctions could make it move from the top of the Tub; and observed a round Bubble of Air to march up, which when once it came to the top, did separate the one from the other. If this hold good, it seems to prove, that neither Mercury, nor any other Liquor would fall down in Pipes, unless there were Air lurking amongst the parts to fill up the deserted space.

From this Experiment we learn, that no person can well apprehend or conceive, how far, and to what bounds the smallest part of Air is able to expand it self. And it proves evidently, that when the *Receiver* is as much emptied as it can be, by the Art of man, yet it is full of Air compleatly.

The

The third trial was after this manner: I set within the *Receiver* a little Glass half full of *Brandy*, and the lid being cemented on, I began to pump, but there appeared no alteration at the first exsuction. At the second, I perceived a great company of very small Bubbles, that for a long time ascended from the body of it, and came to the surface. At the third, they were so frequent, and great, that the *Brandy* appeared to seeth and boil, and by reason of the great ebullitions, much of it ran over the lips of the Glass, and fell into the bottom of the *Receiver*. This boiling continued for the space of 7 or 8 exsuctions, and by process of time, the Bubbles grew fewer and fewer, and when about 30 or 40 exsuctions were made, no more appeared. With this same sort of *Brandy*, I filled the fore-named Pipe, and set it within the *Receiver*, the mouth of the Tub being guarded with the same sort of Liquor. When it began to subside, there appeared no Bubbles near the top as before: the reason seems to be, because the *Brandy* was well exhausted from its *aërial particles*. For a fourth trial, I filled the same Tub with Ale, that was only 5 or 6 dayes old, and drowning the end of it among stagnant Ale of the same kind, I began to Pump, and found, that assoon as the Liquor began to subside, from the top of the Pipe, the whole Ale within the Pipe, almost turned into Air, and Froth, and so many large Bubbles came up from the stagnant Liquor, that I thought the whole was converted into Air. It was most pleasant to behold their several forms and shapes, their order and motion. This same Tub being filled with *sweet milk*, I found very few Bubbles in it, when by the exsuctions, it began to subside. I likewise took a little Glass-viol, and fill'd the half of it full with common Ale, and set it within the *Receiver*. At the first exsuction,

Bubbles

Bubbles of Air began to rife out of it. At the fecond and third, they did fo multiply, that they fill'd the other half of the Glafs, and ran over, as a Pot doth when it boileth. And before I could exhauft all the Air out of it, moe than 20 exfuctions paffed.

For a fifth trial, I filled the often mentioned Pipe with Fountain-water, and when it began to fubfide by Pumping, I found it leave much Air behind it. But all the exfuctions I made, could not make the Water of the Pipe go fo low, as the ftagnant Water, by which impediment, I could Pump no Air out of the Pipe, as I did, while I made ufe of *Brandy*. This tells us, that either there is not fo much Air lurking among *Water*, as among *Brandy*, or that the Air among this, hath a more expanfive faculty in it, than the Air that lurks among *Water*. If any think, that it is not true and real Air, which comes from the *Brandy*, but rather the *Spirits* of it, which evaporats. I anfwer, if a man taft this *Brandy* that's exhaufted of its *aërial particles*, he will find it as ftrong, as before, which could not be, if the *Spirits* were gone,

For a fixth trial, I took a *Frog* and inclofed her within the *Receiver*. But all the exfuctions I was able to make, could not fo much as trouble her. Only, when the *Receiver* was exhaufted, I perceived her fides to fwell very big, and when the *Stop-cock* was turned, to let in the Air again, her fides clapped clofs together. I obferved likewife, when the Air was pretty well Pumped out, that the *Frog* had no refpirations, or if there were any, they were very infenfible. The next day, after fhe had been prifoner in the *Receiver* 24 hours, I began again to Pump, and after feveral exfuctions, her fides fwell'd pretty great, and I perceived her open her mouth wide, and fomewhat like a
Bag

Bag endeavouring to come out, which surely hath been some of her noble parts, striving to dilate themselves, the body being freed of all Pressure from the ambient Air.

OBSERVATION XI.

Take a slender chord, about 4 or 5 yards in length, and fasten the middle of it to the seiling of a Room with a nail, so that the two ends of it may hang down equally. Take next a piece of Wood, two or three foot long, two inches broad, and one inch thick, and boring an hole in each end of it, put through the two ends of the chord, and fasten them with knots; but so, that the piece of Wood may ly Horizontal, and be in a manner a *Pendulum* to swing from the one end of the Chamber to the other. Take next a Bullet of Lead or Iron, about 20 or 24. ounces, and lay it upon the said piece of Wood: but because it cannot well ly, without falling off, therefore nail upon the ends, and the sides of the Timber, four pieces of Sticks, on each end one, and on each side one, as *Ledgets*, for keeping the Bullet from falling off. All things being thus ordered, draw up the piece of Wood towards the one side of the Room, by which means losing its horizontal position, it will ly declining-wise, like the roof of an house. In this position, lay the Iron Bullet in the upmost end of it, and then let them both pass from your fingers, the one end of the Wood going foremost, and you will find it swing towards the other side of the house, and return again, as a *Pendulum*. This motion, if the Wood be well guided in its vibrations, will last perpetually, because in its moving down, the Bullet is hurled from the one end of the Wood, to the other, and hits it so smartly, that it begets in it,

an

an impulse, whereby it is carried farder up, than it would be, without it. By this means, the *vibrations* get not liberty to diminish, but all of them are kept of the same length. In the second vibration, the same Bullet is hurled back again to the other end, and hiting it with all its weight, creats a second impulse, wherewith the Wood is carried, as far up as the point it was first demitted from.

Though this may seem a pretty device to please the fancy, that's many times deceived, while things are presented to it, by way of speculation, yet upon tryal and experience, there will be found, an unspeakeable difficulty: and it's such an one, that a man would not readily think upon. I said, that when the Wood was let go, and was in passing down, the Bullet in it, would hurl down, and hit the oppsite end, and beget an impulse; but there is no such thing, for verily, though the Bullet be laid upon a very declining plain Board, whereupon no man could imagine a round body could ly, yet all the time the Board is in swinging, from the one side of the Chamber, to the other, and consequently, sometimes under an horizontal, and somtimes under an declining position, the Bullet lies dead in the place, where you first placed it. This Observation is not so much for a perpetual motion, as for finding out the reason of this pretty *Phenomenon*, namely, what's the cause, why the Bullet, that cannot ly upon a reclining Board, while it's without motion, shall now ly upon it, while it's under motion? What is more difficult, and nice, to ly upon any thing, that declines from a levell, than *Quick-silver*; yet lay never so much of it upon this Board, while it is swinging, it shall ly dead, and without motion. But no sooner you stop the motion of the wood,

F f   but

but affoon, the Bullet, or the *Quick-filver*, is hurled, either this way, or that way.

OBSERVATON XII.

I Find it mentioned by fome learned perfons, that when a Ship is under Sail, if a ftone be demitted from the top of the Maft, it will move down in a line parallel with it, and fall at the root. Some might think, it ought not to fall directly above the place it hang over, but rather fome diftance behind, feing the Ship hath advanced fo much bounds, in the time, wherein the ftone is coming down. Likewife, while a Ship is under Sail, let a man throw up a ftone never fo high, and never fo perpendicular, as to his apprehenfion, yet it will fall down directly upon his head again, notwithftanding that the Ship hath run (perhaps) her own length in the time, while the ftone was afcending and defcending. This experiment I find to hold true, which may be eafily tryed, efpecially when a man is carried in a Boat upon fmooth Water, drawn by a horfe, as is done in fome places abroad. Let him therefore throw up a little Stone, or any heavy Body, and he will find it defcend juft upon his head, notwithftanding that the Horfe that draggs the Boat, be under a gallop, and by this means hath advanced ten or twelve paces in the time. Or while the Boat is thus running, let a man throw a ftone towards the brink of the VVater; in this cafe he fhall not hit the place he aimed at, but fome other place more forward. This lets us fee, that when a Gun is fired in a Ship under Sail, the Bullet cannot hit the place it was directed to. Neither can a man riding with a full Career, and fhooting a Piftol, hit the perfon he aims at, but muft furely mifs him,

him, notwithſtanding, that though in the very inſtant of time wherein he fires, the mouth of the Piſtol was moſt juſtly directed. For remedy whereof, allowance muſt be granted in the aiming at the mark.

VVhile a man throws up a ſtone in a Ship under Sail, it muſt receive two diſtinct impulſes, one from the hand, whereby it is carried upward, the other from the Ship, whereby it is carried forward. By this means, the ſtone in going up, and coming down, cannot deſcribe a perpendicular, but a crooked Line, either a *Parabola,* or a Line very like unto it. Neither can it deſcribe a perpendicular Line, in coming down from the top of the Maſt, though in appearance it ſeem to do ſo, but a crooked one, which in effect muſt be the half of that, which it deſcribes in going up, and coming down. For this ſame cauſe, a ſtone thrown *horizontally,* or towards the brink of the VVater, muſt deſcribe a crooked Line alſo. And a *Piſtol Bullet* ſhot, while a man is riding at a full Carreer, muſt deſcribe a Line of the ſame kind. Note, that a man walking from the *Stern* of a Ship to the *Head,* walks a longer way, than in walking from the *Head* to the *Stern.* Secondly, a man may walk from the *Head* to the *Stern,* and yet not change his place. 'Tis obſervable, that a man *under board,* will not perceive whether the Ship be ſailing, or not, and cannot know when her *Head* goes about. And it is ſtrange, that when a man is incloſed in a *Hogs-head,* though he have light with him, yet let him be never ſo oft whirled about, he ſhall not know, whether he be going about, or not.

OBSER-

OBSERVATION XIII.

I Found in a *Philofophical tranfaction* lately Printed, that *Decemb.* 13. *1669,* one *Doctor Beal* found the Mercury in the *Barofcope*, never to be fo high, as it was then. That fame very day, I found the hight of it 29 inches, and nine ten parts, which I never obferved before. And though the day here was dark, and the Heavens covered with Clouds, yet no rain for many dayes followed, but much drynefs, and fair weather. On *Saturday* night, *March* 26, 1670, I found the altitude no more than 27, and nine ten parts. This night was exceeding windy, with a great rain. On *February* 1. 1671. I found the altitude 30 inches, and the Heavens moft clear. But in the moft part of *May* following, I have found the hight but 27 inches, and five ten parts, in which time there was abundance of rain.

OBSERVATION XIV.

NOvember 7. 1670. I made exact trial, with the *Magnetick Needle* for knowing the variation, and I found it vary from the *North*, three degrees and a half, towards the Weft. *Hevelius* writes from *Dantzick* to the *Royal Society* at *London, July* 5. 1670, that it varies with him feven degrees twenty minuts, weft.

OBSERVATION XV.

DEcember. 17. 1669, I obferved with a large *Quadrant*, half 9 a clock at night, the formoft *Guard-ftar*, when it was in the *Meridian*, and loweft, to have 41 de-

grees 22 minuts of altitude. And on *January* 7. 1670 at
7 a clock in the morning, I found it, when it was in the *Meridian*, and higheſt, to have 70 degrees, 27 minuts. Hence
I conclude the *elevation of the Pole* here to be 55 degrees,
54 minuts, 30 ſeconds: and conſequently as much at
Edinburgh; becauſe both the places are upon one and the
ſame Parallel.

OBSERVATION XVI.

FOr finding the true *Meridian*, follow this method. In
ſome convenient place fix two Wyre ſtrings with
weights at them, that they may hang perpendicular. Then
in the night time, obſerve, when the *fourth ſtar of the Plough*
begins to come near to the loweſt part of the *Meridian*, at
which time you will find the *Polar ſtar* higheſt. Then, ſo
order the two ſtrings, by moving them hither, and thither,
till both of them cover both the ſaid *Stars*, then ſhall they
in that poſition give you the true *South* and *North*. This
obſervation is the product of the ſevenrh.

OBSERVATION XVII.

THere fell out in *Mid* and *Eaſt-Lothian*, on *Thurſday*
May 11, 1671, in the afternoon, a conſiderable
ſhour of *hail*, with thunder and rain. It came from the
South-weſt, with a great blaſt of wind, and ran alongs from
Pictſ-land-hills North-eaſt, towards the Sea-coaſt. The
hail were big in ſeveral places, as *Muſquet Ball*, and many of
them rather *oval* than *round*. Some perſons ſuffered great
loſs of their *young Peaſe*; others of their *Glaſs Windows*.
Eight or ten days before, there was a conſiderable heat,
and

and dry VVeather. For 20 dayes after , cold Easterly winds, with rain every day, but especially, in the end of the Moneth, extraordinary rain and mist This is so much the more to be observed, because in this Countrey, seldom such extraordinary *hail* falls out. This year the *Agues* and *Trembling Fevers* have been most frequent, and to many deadly.

OBSERVATION XVIII·

I Did hear lately of a curious Experiment in *Germany*, made by a Person of note, which I shall briefly in this Observation, let the *Reader* understand. And though I have heard since, that it is now published in Print, yet I hope it will not be impertinent to mention it here, especially for their cause, who cannot conveniently come to the knowledge of such things. And for this reason also, that I may explicat the *Phenomena* thereof, from the foregoing doctrine , and demonstrat particularly the true cause of that admirable effect , that's seen in it, which I desiderat in the publisher. The Auctor then takes two *Vessels* of *Brass*, each one of them in form of half a sphere, of a pretty large size. Nothing can more fitly represent them for form and quantity, than two *Bee-skeps*. Only, each of them, hath a strong *Ring* of Brass upon the *Center* without : and they are so contrived by the *Artist*, that their orifices agree most exactly, so that when they are united, they represent an intire Sphere almost. In one of the sides, there's a hole, and a *Brass Spigot* in it, through which the whole *Air* within, is exsucted, and drawn out, namely by the help of the *Air-pump*. And, when by several exsuctions the *Vessels* are made empty, the *Stop-cock* is turned about,

about, by which means, no *Air* can come in. And, they
remaining empty, are taken from the *Pump*, and do cleave
so fast together, that though a number of *lusty fellows*, 12
on each side, do pull vigorously, by help of ropes fastned
to the *Rings*, yet are they not able to pull them asunder.
And because this will not do it, he yokes in 12 *Coach
Horses*, six on every side, yet are they not sufficient, though
they pull contrariwise to other, to make a separation. But
to let the *Spectators* see, that they may be pulled asunder,
he yokes in 9 or 10 on every side, and then after much
whipping, and sweating, they pull the one from the
other.

The cause of this admirable effect, is not the fear of *vacuity*, as some do fancy, for if that were, all the *Horses* in
Germany would not pull them asunder, no not the strength
of *Angels*. It must then be some extrinsick weight and
force, that keeps them together, which can be nothing
else, but the weight of the *invironing Air*. Because, no
sooner a force is applied, that's more powerful, than the
weight of the *Air*, but assoon they come asunder. And
so neither six men, nor six horses on each side are able to
do it : but nine or ten on each side makes a separation. For
understanding the true cause of this *Phenomenon*, we must
consider that the *Vessels* are 18 inches in *diameter*. If this
be, then according to the last Experiment, there are two
Pillars of *Air*, each one of them as heavy as a Pillar of *Mercury* 18 inches thick, and 29 inches long, by which they
are united. Or, each Pillar of *Air*, is as heavy, as a Pillar
of *Water* 34 foot high, and 18 inches in *diameter*. For
finding the weight of it in pounds, and consequently, the
weight of each Pillar of *Air*, by which the two *Vessels* are
united, follow this method. First, multiply 9 the semi-
diameter

diameter of the *Pillar*, by 54 the circumference, and this
gives you 486,the half whereof is the bounds of the *Area*,
namely 243. And becaufe 34 foot contains 408 inches,
I multiply 408 by 243,the product whereof is 99144; fo
many fquare inches are in a Pillar of Water 34 foot high,
and 18 inches thick, Now feing there are 1728 inches
in a *cubical foot*, I divide the number 99144, by this number, and I find 57 fquare foot of Water, and more. And
becaufe every fquare foot weighs 56 pound *Trois*, I multiply 56 by the number 57, and the product is 3192 pound,
which is the juft weight of a Pillar of *Water* 34 foot high,
and 18 inches in diameter, and which is the juft weight alfo of each Pillar of *Air*, by which the two *Veffels* are kept
together , which will be more weight than *feven Hogs-
heads full of Water*. This is eafily known; for feing a quart
of our meafure weighs feven pound, (or to fpeak ftrictly
fix pound fourteen ounces , feing the *Standard-jug* of *Stri-
viling* contains three pound feven ounces of Water) a gallon muft weigh 28 pound: but 16 times 28, is 448. A
Puncheon then full of Water, weighs 448 pound. If then
you divide 3192 by 448, you will find more than 7. The
9 horfes then upon this fide have 3192 pound weight to
draw, or 199 ftone, or the weight of *feven Hogs-heads
full of Water*. The other 9 horfes upon the other fide, have
as much to pull, 'Tis no wonder then to fee fo much difficulty and pains to make a feparation. It is obferved, that before the *Air* be exfucted and drawn out of the two *Veffels*,
one man is able to pull them afunder with his hands only.
Nay, which is more, if he but blow into them, as a man doth
into a Bladder, he will feparat them, The reafon is, becaufe the *Air* within, is of as great force, as the *Air* without. 'Tis obfervable next, that the larger the *Veffels* be

in *diameter*, the more strength is required to pull the one from the other. Upon suppofition then, they were 4 foot wide, I verily believe 30 yoke of oxen, upon every fide, would hardly difjoyn them; becaufe the weight of each Pillar of Air, would be no lefs, than 22844 pound, which would take 63 ftrong horfes to overcome the force of it. To pull the one *Veffel* therefore from the other, there muft be 126 horfes, that is, 63 on every fide.

OBSERVATION XIX.

THough thisObfervation may feem ufelefs, becaufe the Propofals, that are mentioned in it, cannot be made out, and brought to pafs, the *Author* having died, before he had encouragment to profecute them : yet for thefe following reafons, I have adventured to infert it here. Firft, that others, may either be minded to find out (if poffible) his inventions, or fet a work to find out fomethings, that may be as ufeful. Next, becaufe, he was one of this fame *Nation*, and a great *Mafter* of the *Mathematicks*, not only in the *Speculative*, but in the *Practical* part chiefly, and admirable for invention. And for this caufe principally I have prefumed to mention his defigns, and propofals, which were found among his *Notes*, after his death, which are here infert, as they were written with his own hand, and offered to the publick ; not only at home, but abroad to ftrangers. There have been men in all ages famous, for fome one Art and Science beyond others, as *Apelles* for Painting, *Hippocrates* for Medicine, *Demofthenes* for Oratry, but who have been more famous in their time than fome perfons for their profound knowledge in *Aftronomy*, *Geometry*, and the other parts of the *Mathematicks*. What

an

an admirable perfon was *Archimedes* for his divine knowledge, both in the *Speculative*, and *Practical* part. Yet,
it was not his fpeculations fimply, though excellent, that
did fo much commend him, as his *Inventions*, and admirable *Engines* for peace and war, as is clear from the *Romane
Hiftories*, and others. I confefs the Students of thefe Arts,
are not fo much in requeft now, at leaft amongft fome, and
that knowledge is not fo much efteemed ; and the reafon
may be ; becaufe fome who profefs themfelves *great Ma-
fters*, ftudy nothing but the pure fpeculations, which fometimes are to fmall purpofe, others before knowing the fame,
unlefs for perfecting of the mind, and giving to a man fome
private fatisfaction. But fuch things will never commend
a man fo much as the practical part, and new Invention will
do. 'Tis furely a fmall bufinefs for one to do nothing, but
to nibble at fome petty Demonftration. But when fuch
fpeculations are joyned with invention and practice, for the
profit, and ufe of men, among whom they live, then are
they far more to be commended. And if this be not, fuch
knowledge is of fmall advantage to themfelves or others.
Many of the Ancient, and late Aftronomers have been, and
are famous for practice, as witnefs the indefatigable pains
they have been at in making their Obfervations. What
hath fo highly commended *Merchifton* over all *Europe*, as
his inventions, efpecially his *Logarithmes* ? And if all be
true, that's reported (which I am apt to believe) he might
have been more renowned, for his many excellent *Engines*,
which though ufeful, yet becaufe hurtful to mankind, he
buried with himfelf. I am confident, if the Author of
thefe propofals had had time to have profecuted them, he
would have been celebrated in the Catalogue of the moft
famous *Mathematicians* of his time. But leaving this, I
shall

shall give you them in his own words: but first his Apology.

These bold proposals will need perhaps an apology to such, to whom the causes, and circumstances are unknown. Let it suffice, that the Proposer finding himself between two extreams, either to leave unprosecuted this affair, for fear of being mistaken by some, as impudent, or to commit himself openly to the charitable judgement of others, who will suspend their censure, till they have seen what his endeavours will produce. He hath rather chosen this last, especially considering, that his silence could not answer to his duty, which he owes to his Countreys service, seing the following Engines may be so useful to it. A deduction of the fabrick, causes, and occasions of these *new Engines*, that set the Inventer a-work, would take a long time to discourse upon. This Paper therefore is only destined for a short information of their use, the rest, which could not here be insert without impertinency, may be supplied afterwards (if need be) either by a discourse, or by a particular demonstration. The Proposer then is of opinion, (if self-love of his own Inventions do not blind his Judgement) that these *paradoxes* may be truly affirmed.

That if it shall please His Majesty to arm with these new Arms, and Engines, 500 Foot, or fewer, this small number shall be Masters of the Fields in *France, Germany, Spain,* or where else it shall please His Majesty, however encountered by the most powerful Army of Horse or Foot, armed with ordinary Arms, of Pistol, Carabine, Pike, Musquet, which Europe can bring to the Fields.

The cause of this admirable effect, is in the quality of these new Arms, by which, the whole Horsemen and Footmen of the enemy are rendred useless, and unservicable;

neither

neither can they do any offence to thefe, who are fo
armed.

The *Mufquetteers*, who can only ferve againft thefe *Ma-
chins*, fhall be put to fuch difadvantage, as it is impoffible
they can ftand, the leaft time, in the common way of fervice with the *Mufquet*, it not being able to make one fhot
for twenty, which fhall be made from thefe new Engines.

Thefe *new Arms*, have this advantage likewife, that
thefe who are fo armed, can by no force of Horfe or Foot
be broken, or put to diforder. The Souldiers are alfo by
them put to a neceffity of keeping together, and fighting,
and by them, they are fo *Baricado'd*, and ftrongly defen-
ded, that if they leave them not, they cannot be expofed
to danger. This contributes much to good Difcipline,
when the Souldiers fhall by neceffity be tied to his duty,
and fear, which otherwife makes him run away, fhall here
for his fafety make him ftand.

Thefe *new Arms* are ufeful, as well in Marching, as in
Combating, for with them, we may march fecurely two
in front, through the ftraiteft paffages, and be able to force
with them any advantage a ftrait paffage can give to an enemy. Befides, for a long hafty march, where Victuals cannot be well carried, the Souldiers are able with thefe *Arms*
to carry their own provifion for eight dayes, with more facility, then they can now carry one dayes provifion.

To lodge in the open fields, thefe *Arms* fhall need no
Intrenching, for they fufficiently both Arm and *Baricade*
the Souldiers.

And as they are ufeful in Service, fo are they a great deal
cheaper than the ordinary *Arms*. For although with
5 thoufand men fo armed, the fervice of 100000 armed
with common Arms may be done, yet the whole price of
                                                them

them will not amount to that which will be required for arming 20500 *Corraſſiers*, as may be particularly deduced, from the particular prices of the Arms, and Engines fitted for the ſervice of 5000 men. The Propoſer doth offer to ſhew, that theſe Arms will not ſurmount 40000 pound *Sterling*. The *Artillery* will amount to 4500, and the payments of this number of men ſo armed, yearly to 70000 pound. Yet all theſe are taken in ſo large a latitude of reckoning, as the ſum of *Arms*, *Artillery*, and payments, will not be much above 130000 pound *Sterling*.

The Arms from which this effect is promiſed, are *new Engines*, with which one man is able to do the ſervice of a great many *Muſquetteers*. And thoſe are of two ſorts, either to be uſed upon a ſmall Wagon for Footmen, or on a greater for a Horſe, with either of which, one hand is able to make the fire of 100 *Muſquetteers*, and ſo much better, by how much it is more regularly, and fitly done for execution and offence. The new Cannon ſhall have the like advantage above the old, both for eaſie carriage, being lighter, and for greater execution, ſhooting ſix, nine, or twelve Bullets for one. Theſe Arms give not only this advantage at Land in the field, but alſo in Ships, and places of defence.

Theſe nine following propoſitions he likewiſe offered to make good,

Firſt, With one ſhot of *Cannon*, to do the execution of five ſhot of the ſame *Cannon*, in the common way of Battery.

Secondly, to diſable any Ship or Galley with one ſhot of *Cannon*.

Thirdly, to fire any combuſtible matter with the ſhot of a *Cannon*.

Fourthly,

Fourthly, to make an *Machin* or *Engine* for tranſporting an Army, which may be carried without the incommodity thereof.

Fifthly, to make a flotting *Fortreſs* for defence of Rivers, and prohibition of Paſſages.

Sixthly, to make a *Mortar* that hath a *directory Stell* upon the Carriage.

Seventhly, to make *Petards* of divers forms, that ſhall be able to do twice as much execution, as thoſe that contain as much Powder.

Eighthly, to make ſmall *Petards* of great effect.

Laſtly, to make *Bridges*, and *Scaling Ladders* of eaſie Carriage.

OBSERVATION XX.

THeſe Obſervations being *Miſcellany*, require not a formal connexion between themſelves, and therefore 'tis no matter what method I keep in ſetting them down. And though this may ſeem not ſo pertinent, as others, yet becauſe the deſign of it is only *Philoſophical* and for advancing the *Hiſtorical* part of Learning in order to *Spirits*, upon which the *Scientifical* part doth ſo much depend, I have preſumed to inſert it here, conſidering alſo that there are ſome, who have adventured to deny their exiſtence. and being; which from ſuch a Hiſtory as this, may be more than probably evicted. I find likewiſe, that ſeveral Writers have remarked ſuch ſtrange accidents, and have tranſmitted them to poſterity , which may ſerve to good uſe. The ſubject-matter then of this Obſervation is a true and ſhort account of a remarkable trial, wherewith the Family of one *Gilbert Compbel*, by Profeſſion a Wea-

ver in the old Paroch of *Glenluce* in *Galloway*, was exer-
cifed. Though the matter be well known to feveral perfons
at that time, and fince too; yet there are others, eighteen
years interveening, to whom (perhaps) fuch a relation
will not be unacceptable, who have either not as yet heard
of it, or at leaft, have not gotten the true information,
which is here fet down, as it was Written, at the defire of
a fpecial Friend, by *Gilbert Campbel*'s own Son, who knew
exactly the matter, and all the circumftances, whofe words
are as follows.

It happened in *October* 1654, that after one *Alexander
Agnew*, a bold and fturdy Beggar, who afterwards was
hanged at *Dumfreis* for blafphemy, had threatned hurt to
the Family, becaufe he had not gotten fuch an alms as he
required : the faid *Gilbert* was oftentimes hindered in the
exercife of his Calling, all his Working-Inftruments being fome of them broken, fome of them cutted, and yet
could not know by what means this hurt was done ; which
piece of trouble did continue, till about the middle of *November*, at which time the Devil came with new and extraordinary affaults, by throwing of Stones in at Doors and
Windows, and down thorow the Chimney-head, which
were of great quantity, and thrown with great force,
yet by *Gods* good providence, there was not one perfon of
the Family hurt, or fuffered dammage thereby. This
piece of new and fore trouble, did neceffitat Mr. *Campbel*
to reveal that to the *Minifter* of the Paroch, and to fome
other Neighbours and Friends, which hitherto he had en-
dured fecretly. Yet notwithftanding of this, his trouble
was enlarged ; for not long after, he found oftentimes his
Warp and Threeds cut, as with a pair of Sizzers, and the
Reed broken : and not only this, but their apparel cut after

ter the ſame manner, even while they were wearing them,
their Coats, Bonnets, Hoſe, Shooes, but could not diſcern how, or by what mean. Only it pleaſed *God* to pre-
ſerve their perſons,that the leaſt harm was not done. Yet,
in the night time, they wanted liberty to ſleep, ſomething
coming, and pulling their Bed-cloaths and Linnings off
them,and leaving their bodies naked. Next, their Cheſts,
and Trunks were opened, and all things in them ſtrawed
here and there. Likewiſe, the parts of the Working In-
ſtruments, that had eſcaped, were carried away, and hid
in holes and bores of the houſe, where hardly they could
be found again. Nay, what-ever piece of Cloath, or
Houſhold-ſtuff, was in any part of the houſe, it was carried
away, and ſo cut and abuſed, that the Good-man was neceſſitated with all haſte and ſpeed, to remove, and to tranſ-
port the reſt to a Neighbours houſe, and he himſelf compelled to quite the exerciſe of his Calling, whereby only
he maintained his Family. Yet, he reſolved to remain in
the houſe for a ſeaſon. During which time, ſome perſons
about, not very judicious, counſelled him to ſend his children out of the Family, here and there, to try whom the
trouble did moſt follow, aſſuring him, that this trouble
was not againſt all the Family, but againſt ſome one perſon, or other in it, whom he too willingly obeyed. Yet,
for the ſpace of four or five dayes after, there were no remarkable aſſaults, as before. The Miniſter hearing thereof, ſhewed him the evil of ſuch a courſe, and aſſured him,
that if he repented not, and called back his children, he
might not expect that his trouble would end in a right way.
The children that were nigh by, being called home, no
trouble followed, till one of his ſons, called *Thomas*, that
was farreſt off, came home. Then did the Devil begin a-
freſh;

fresh; for upon the *Lords Day* following, in the afternoon, the house was set on fire, but by his providence, and
the help of some people, going home from Sermon, the
fire was extinguished, and the house saved, not much loss
being done. And the *Monday* after, being spent in pri-
vat Prayer and Fasting, the house was again set on fire
upon the *Tuesday* about nine a Clock in the morning, yet
by providence, and the help of Neighbours, it was saved,
before any harm was done.

Mr. *Campbel*, being thus wearied, and vexed, both in
the day, and in the night time, went to the *Minister*, de-
siring him, to let his son *Thomas* abide with him for a time,
who condescended, but withal assured him, that he would
find himself deceived, and so it came to pass: for, notwithstanding that the child was without the family, yet were
they, that remained in it, sore troubled both in the day
time, and in the night season, so that they were forced to
wake till mid-night, and sometimes all the night over.
During which time, the persons within the Family, suffer-
ed many losses, as the cutting of their Cloaths, the throwing of Peits, the pulling down of Turff, and Feal from the
Roof, and Walls of the House, and the stealing of their
Apparel, and the pricking of their flesh and skin with Pins.
The *Presbytery* having conveened at the place, for a solemn
Humiliation, perswaded *Gilbert Campbel* to call back his
Son *Thomas*, notwithstanding of whatsoever hazard might
follow. The Boy returning home, affirmed that he heard
a voice speak to him, forbidding him to enter within the
house, or into any other place where his Fathers Calling
was exercised. Yet he entered, but was sore abused, till
he was forced to return to the *Ministers* house again.

Upon *Monday* the 12 of *February*, the rest of the Family

ly began to hear a voice speak to them, but could not well know from whence it came. Yet, from evening till midnight, too much vain discourse was kept up with the _Devil_, and many idle and impertinent questions proposed, without that due fear of _God_, that should have been upon their Spirits, under so rare and extraordinary a trial. The _Minister_ hearing of this, went to the house upon the _Tuesday_, being accompanied with some Gentle-men, who after Prayer was ended, heard a voice speaking out of the ground, from under a bed, in the proper Countrey Dialect, saying, _Would ye know the_ Witches _of_ Glenluce _? I will tell you them_; and so related four or five persons names, that went under an evil report. The said _Gilbert_ informed the company, _That one of them was dead long ago._ The _Devil_ answered, and said, _It is true, she is dead long ago, yet her spirit is living with us in the world._ The Minister replied, saying, (though it was not convenient to speak to such a person) _The Lord rebuke thee Satan, and put thee to silence; we are not to receive any information from thee, whatsoever fame any persons go under. Thou art but seeking to seduce this Family: for_ Satans _Kingdom is not divided against it self._ After which all went to Prayer again, which being ended (for during the time of Prayer no trouble was made) the _Devil_ with many threatnings boasted and terrified the Lad _Thomas_, who had come back that day with the _Minister_, that if he did not depart out of the house, he would set all on fire. The _Minister_ answered, and said, _The Lord will preserve the House, and the Boy too, seing he is one of the Family, and hath Gods warrand to tarry in it._ The _Devil_ answered, _He shall not get liberty to stay: he was once put out already, and shall not abide here; though I should pursue him to the end of the world._ The _Minister_ replied, _The_

Lord

Lord will stop thy malice against him. And then they all prayed again, which being ended, the *Devil* said, *Give me a Spade and a Shovel, and depart from the house for seven dayes, and I shall make a grave, and ly down in it, and shall trouble you no more.* The *Good-man* answered, *Not so much as a Straw shall be given thee, through Gods assistance, even though that would do it.* The *Minister* also added, *God shall remove thee in due time.* The *Devil* answered, *I will not remove for you, I have my Commission from Christ to tarry, and vex this Family.* The *Minister* answered, *A permission thou hast indeed, but God will stop it in due time.* The *Devil* replied, *I have* (Mes. *John*) *a Commission, that (perhaps) will last longer than your own.* After which, the *Minister* and the *Gentlemen* arose, and went to the place where the voice seemed to come from, to try if they could find any thing. And after diligent search, nothing being found, the *Gentlemen* began to say, *We think this voice speaks out of the children,* for some of them were in their beds. The *Devil* answered, *You lie, God shall judge you for your lying, and I and my Father will come and fetch you to hell, with* Warlock-theeves; and so the *Devil* discharged the *Gentlemen* to speak any, saying, *Let him speak that hath a Commission* (meaning the *Minister*) *for he is the Servant of God.* The *Gentlemen* returning back with the *Minister*, they sat down near to the place whence the voice seemed to come from ; and he opening his mouth, spake to them, after this manner. *The Lord will rebuke this Spirit, in his own time, and cast it out.* The *Devil* answering, said, *It is written in the* 9 of Mark, *the Disciples could not cast him out.* The *Minister* replied, *What the Disciples could not do, yet the Lord having hightned the Parents faith, for his own glory did cast him out, and so shall*

he

he thee. The *Devil* replied, *It is written in the* 4 *of* Luke, *And he departed, and left him for a ſeaſon.* The *Miniſter* ſaid, *The Lord in the dayes of his humiliation, not only got the victory over* Satan, *in that aſſault in the wilderneſs, but when he came again, his ſucceſs was no better, for it is written,* Joh. 14. *Behold the Prince of this world cometh, and hath nothing in me ; and being now in glory, he will fulfill his promiſe, and God ſhall bruiſe* Satan *under your feet ſhortly,* Rom. 16. The *Devil* anſwered, *It is written,* Mat. 25. *There were ten Virgins, five wiſe, and five fooliſh ; and the Bridegroom came: The fooliſh Virgins had no Oyl in their Lamps, and they went unto the wiſe to ſeek Oyl; and the wiſe ſaid, Go and buy for your ſelves : and while they went, the Bridegroom came, and entered in, and the door was ſhut, and the fooliſh Virgins were ſent to hells fire.* The *Miniſter* anſwered, *The Lord knows the ſincerity of his ſervants, and though there be ſin and folly in us here, yet there is a fountain opened to the houſe of* David *for ſin and for uncleanneſs, and when he hath waſhed us there, and pardoned all our ſins, for his Names ſake, he will caſt the unclean ſpirit out of the land.* The *Devil* anſwered and ſaid, *That place of Scripture is written in the* 13 *of* Zechariah, *In that day I will cauſe the Prophets, and the unclean ſpirit, paſs out of the land; but afterwards it is written, I will ſmite the Shepherd, and the Sheep ſhall be ſcattered.* The *Miniſter* anſwered and ſaid, *Well are we, that our bleſſed Shepherd was ſmitten, and thereby hath bruiſed thy head, and albeit in the hour of his ſufferings, his Diſciples forſook him,* Mat. 26. *yet now having aſcended on high, he ſits in glory, and is preſerving, gathering in, and turning his hand upon his little ones, and will ſave his poor ones in this Family from thy malice.* The *Miniſter* returning back a little, and ſtanding upon the floor,

the

the *Devil* said, *I knew not these Scriptures, till my Father taught me them.* I am an *evil Spirit, and Satan is my Father, and I am come to vex this house* ; and presently there appeared a naked hand, and an arm, from the elbow down, beating upon the floor, till the house did shake again ; and also the *Devil* uttered a most fearful and loud cry, saying, *Come up Father, come up : I will send my father among you. See, there he is behind your backs.* The *Minister* said, *I saw indeed an hand, and an arm, when the stroak was given, and heard.* The *Devil* said to him, *Saw you that ? It was not my hand, it was my fathers ; my hand is more black in the loof. Would you see me ? Put out the candle then, and I shall come butt the house among you like fire-balls.* After which all went to Prayer, during which time, it did no harm, neither at any other time when *God* was worshipped. When Prayer was ended, the *Devil* answered and said, *Mes John, if the Good-mans sons prayers at the Colledge of Glasgow, did not prevail more with God, than yours, my father and I had wrought a mischief here ere now.* To which one of the Gentlemen replied, though a check had been given him before, *Well well, I see you confess there is a God, and that prayer prevails with him, and therefore we must pray to God, and will commit the event to him.* To which the *Devil* replied, *Yea Sir, you speak of prayer, with your broad lipped Hat* (for the Gentleman had lately gotten a new Hat in the fashion with broad lips) *I'le bring a pair of Shears from my father, that shall clip the lips of it a little.*

The night now being far spent, it was thought fit every one should withdraw to his own home. Then did the *Devil* cry out fearfully, *Let not the Minister go home, I shall burn the house if he go* ; and many other wayes did he

threaten

threaten. And after the *Minifter* was gone forth, the *Good-man* being inftant with him to tarry, whereupon he returned, all the reft of the company going home. Then faid the *Devil* to the *Minifter*, *You have done my bidding. Not thine*, anfwered he, *but in obedience to God, have I returned to bear this man company, whom thou doft afflict.* Then did the *Minifter* call upon the Name of *God*, and when Prayer was ended, he difcharged Mr. *Campbel*, and all the perfons of the Family, from opening their mouth, in one word to the *evil fpirit*, and when it fpake, that they fhould only kneel down, and fpeak to God. The *Devil* then roared mightily, and cryed out, *What? Will ye not fpeak to me? I fhall burn the houfe, I fhall ftrike the bairns, and do all manner of mifchief.* But after that time, no anfwer was made to it, and fo for a long time no fpeech was heard. After this, the faid *Gilbert* fuffered much lofs, and had many fad nights, not two nights in one week free; and thus it continued till *April*. From *April* to *July*, he had fome refpite, and eafe. But after, he was molefted with new affaults: and even their Victuals were fo abufed, that the Family was in hazard of ftarving; and that which they did eat, gave them not the ordinary fatisfaction they were wont to find.

In this fore and fad affliction, Mr. *Campbel* refolved to make his addrefs to the *Synod* of *Presbyters*, for advice and counfel what to do, which was appointed to conveen in *October* 1655, namely whether to forfake the houfe and place, or not? The *Synod* by their Committee, appointed to meet at *Glenluce* in *Feb.* 1656, thought fit, that a folemn Humiliation fhould be kept thorow all the bounds of the *Synod*, and amongft other caufes, to requeft *God* in behalf of that poor afflicted Family, which being carefully

done

done, the event was, through the Prayers of his People, that his trouble grew less till *April*, and from *April* to *August*, he was altogether free. About which time, the *Devil* began with new assaults, and taking the ready meat that was in the house, did sometimes hide it in holes by the door-posts, and at other times did hide it under the beds, and sometimes among the Bed-cloaths, and under the Linnings ; and at last, did carry it quite away, till nothing was left there, save Bread and Water to live by. After this, he exercised his malice and cruelty against all the persons of the Family, in wearying them in the night time, with stirring and moving thorow the house, so that they had no rest for noise, which continued all the moneth of *August* after this manner. After which time, the *Devil* grew yet worse, and began with terrible roarings, and terrifying voices, so that no person could sleep in the house, in the night time, and sometimes did vex them with casting of stones, striking them with staves on their beds in the night time : and upon the 18 of *September*, about midnight, he cried out with a loud voice, *I shall burn the house* ; and about three or four nights after, he set one of the beds on fire ; which was soon extinguished, without any prejudice, except the bed it self : and so he continued to vex them.

OBSER-

OBSERVATION XXI.

I Need not make any apology for inserting this Observation, even though it be well known upon the matter in this place. But because the thing is extraordinary, and that there are many who have not so much as heard of it, I have therefore presumed to mention it here. The matter is shortly this. There's a certain Woman, named *Mistris Low*, who had a real and true Horn, growing upon the right side of her Head, three inches above her right Ear. The length of it is eleven inches, and two inches about. The form is crooked spirally. It is convex on the outer side, and somewhat guttered in the inner side. It is hard and solid, and all very near of the same greatness. It is not hollow within, as horns are ordinarily, but full, yet it seems to be spongious as a Cane is. It was seven years in growing, and was cut off in *May* 1671, by *Mr. Temple*, an expert Chirurgeon here at *Edinburgh*.

OBSERVATION XXII.

THis Observation is for finding the *Primum vivens* in *Animals*. Albeit I doubt not but the *red Spirit*, or *Blood*, in most *Terrestrial Animals*, is the first product of the *Primigenial juice*, and therefore not improperly named the true *Callidum Innatum* of these Creatures, by the Noble and Ingenious *Harvey*, in his Book *de Generatione*. Neither do I scruple to yeeld, that the *Heart*, and appendent Vessels, are the first formed, and perfected parts in the hotter kind of *Animals*: yet I am confident to affirm, that in many of the colder, and moister kinds of *Aquaticks*, if not in all, neither the redness and heat of the
Vital

Vital Spirits, nor the formation of the *Heart*, *Liver*, &c. are previously requisite, to the structure and existence of the other parts; seing the light of life, which at first inhabited the clear and Criftalin radical moisture, before the formation of any particular part, doth alwayes move in every living creature, according to their particular exigency, without any absolute dependency upon any one part, or member (excepting singular conditions, wherein they may be stated) as to its substance, light, and motion: there being in some *Animals* a simple undulation, in others a flow creeping, but in the more perfect, an impetuous running, or rather flying of the *Vital Spirits*, necessarily required for illumination and vivification of the whole.

For confirmation, I shall give you this singular Experiment. About the middle of *March*, the sperm of *Frogs* (according to the number of *Prolifick Eggs* therein contained) sends forth a multitude of small round Creatures, covered with a black, and moveable *Frock*, which about the end of *March*, and beginning of *April*, by the Gyrations of a Tail behind, like a Rudder, do slowly move their bodies in the Water. At this time having opened severals of them, I found nothing (apparent to the naked eye) but a clear thin Membran, under the fore-named black *Frock*, within which were contained a clear Water, and some small Fibres like Intestines, and in the fore-part a small orifice like a mouth. About the middle of *April*, its motion is more vigorous, and the *Tripes* within are most evident, lying in a very fine circular order, but as yet, there is no Vestige of *Heart*, *Blood*, or *Liver*, &c. About the middle of *May*, the feet formed like small threeds, appear thorow the black *Coat*: within the Breast, the Heart is then visible, of a white and Fibrous substance, the Liver

is

is white, and the Gall therein eafily difcerned. But (which is the head of this Experiment) the Vital Spirit, in form of a clear and pure Water, is manifeftly received by the Nervous Heart, and by the contraction thereof tranfmitted to all the Body, thorow white tranfparent Veffels, which being full of this Liquor, do reprefent the *Lymphatick*, rather than the *Sanguiferous Veins*. Laft of all do the *Pneumatick Veficles* (which in this *Amphibium* fupply the place of the Lungs) arife in the Breaft, after whofe production, the Lympid and Cryftalin Liquor, while the Heart is turgid therewith, feems to be red and fiery, but in the other Veffels, it is of a faint pale colour, untill (about, or near the end of *June*) the *Frock* being caft off, and a perfect *Frog* formed, the whole Veffels are full of Blood, or a red fubftance very thin, and clear: the Liver, and *Pneumatick Veficles*, &c. become red, and Rofy; fo that the Blood in this *Amphibium* (which in the more perfect Animals is firft compleat) feems to be the laft part in attaining its perfection.

That *Salmonds*, and great *Trouts* have an aqueous liquor which runs thorow their *Arteries*, and *Veins*, before their Blood attain the true confiftency, and faturat tincture I am certain: whether it hold in many others, I fufpect, but dar not affirm. Hence it may be (if mens obfervations, were frequent in all kind of *Anatomical infpections*, in feveral *Embryo's* of every *fpecies*) it would be found evident, that the Blood in all thefe, called 'ιιάιμχι hath its immediat original from a fimple homogeneous, and uniform liquor , and doth by gradual and frequent influences of the vital ferment of the heart, receive at length the full tincture, effence, and fubfiftence requifite for vivification , and illumination of the whole members.

Whether

Whether this Experiment doth not fufficiently impugn the univerfality of the hearts firft living, the original of the Gall from the fervour, and ebullition of the Blood, the production of the Blood by the Liver, and many other ancient errors, let any judge, who will but take pains to make and compare *Harveys* trials *de ovo*, with this of the *Porwigl* or *Gyrinus*, *ab ovo*.

Yea, if the aqueous liquor, be not one with the vital Spirit, and fubfequent Blood, then my eyes, and tafte are altogether erroneous.

Moreover, it were to be wifhed, that *Phyfitians* would not fimply ftand upon the *Galenick fuppofitions* of the four alledged Components of the Blood, nor any fuch, or equivalent fancies of the latter *Chymifts*; but that they would ferioufly examine the firft original, and rife thereof from the *Primigenial juice*, or *liquamen*, the progrefs, and perfection of its tinctures, how many renovations, or new tinctures it is capable of; the vaft difference between the Blood of old and young *Animals*, (though, it may be, they are both *univocal* fubftances, while in their integrity within the Veffels) with the *fpecifick* difcriminations, not only of that of any one *Aquatick*, from any *Volatil*, or *Terreftrial*, but likewife of any one *Species* living in the fame Element, with thefe that enjoy the fame Aliments, but of a different *Species*. And laftly, the variety of particular conftitutions, and fingular properties of individual *Animals*, radicated in the fountain of life, or firft original of the Blood. If thefe things, and many more, were truly inquired after (though the Cook be fometimes neceffitated to throw away fome of the Broth with the Scum) I doubt not but the *Neoterick* Invention of *Transfufion of Blood*, would prove altogether ridiculous, and the ancient miftake of too

much

much *Profusion* of this treasure by *Phlebotomy*, might suffer some reasonable checks from infallible Experience, and found reasons, not here to be mentioned. There are truths in Natural Philosophy, which (I doubt not) but found reason and experience will convince the vain world of in due time.

OBSERVATION XXIII.

THis Observation is concerning the aliment and growth of Plants. The inquisitive wits of this, and the last age, having rejected the old opinion of the earths nourishing of Plants, or being converted into their aliment, have made many laudable Experiments for finding out the materials, and means of their growth, and vegetation, such as Sir *Francis Bacon's Observe of Germination, Helmonts* of a *Willow*, and the Noble *Mr. Boyl's* of a *Gourd*, &c. For though a Tree be cut down, and the root thereof wax old in the earth, and the stock die in the ground, yet through the sent of Water, it will bud, as *Job* speaketh, *Chap.* 14. 7, 8, 9. I shall add a short remark of a Willow growing without earth. Upon the 13 of *April* 1662, I set a top branch of the *Peach-leaf'd Willow* in a Glass-viol, among 12 ounces of pure Spring Water, with three small buds upon the top thereof, scarce yet discernable. The first ten or twelve dayes, little white specks appeared upon the sides of the Willow, like small drops of *Quick-silver*, or like the first Bubbles that arise upon the fermentation of Ale or Wine, but no consumption of the Water all this time. Indeed the *Gemms*, which stood three inches above the Water, did visibly swell about the twelfth day. About the fifteenth day, I perceived small white roots within the
Water,

Water, upon several places of the *Plant*, and observed the
Liquor grow somewhat thick, and decay in bulk considerably. Having perceived this, I took another Glass of the
same bigness, with that wherein the *Willow* grew, and
having filled both top-full with Spring Water, I observed clearly the consumption of the Water wherein the
Plant stood, to be so great, that during *May*, *June*, and
a great part of *July*, every week (at least) an ounce and
an half, or two ounces of it were insensibly spent : whereas
the other Water, standing by in an open Vessel of the same
size, made not waste of one spoonful in a whole moneth.
About the middle of *August*, the Water turned very thick,
and green, like that whereon *Duck-weed* useth to grow,
and the fair white roots were all obscured from the fight,
although the Vessel by the multitude of roots was not capable of the third part of Water it received at first. At
this time the branches were advanced to half the bigness,
and a much greater length, than the whole stock, at its
first planting ; and the leaves of as fresh a verdure, as any
Willow in the fields. Thus, having observed, that a tree
of four ounces weight, could in three moneths time, and
little more, consume insensibly, seven or eight times its
own weight of pure Water, without the warm preservation of the earth, and by its own proper digestion, to thicken
the remnant of the Water, that it might serve for *lorication* of the tender fibres of the roots, I took the Glass, the
Tree, and all, and threw them over a Window, supposing
it needless to recruit the Water any more, and judging it
impossible without the warm guard of the earth, that the
naked Tree could be preserved in Winter : yet it had the
good fortune to fall among some thick Herbs in the corner
of a little Garden, where (after it had lien all Winter) it
was

was found, and brought back to me, the branches fairly budding in *April*, the whole Tree frefh and green, yet very little Water was left in the Glafs, by reafon, as I judged, it had fallen upon its fide. Then I endeavoured to keep Water about it, but the Stock filling the neck of the Viol, and the Roots the whole body thereof, the ftarved *Plant* died in *May*, after it had lived a whole year without earth. From this it would feem, that this kind of Tree, (and it may be, many moe) doth diffipat infenfibly fix times more Liquor, than it doth affimilat, and by confequence, that a great quantity of moifture is neceffary for maintainance of great Woods. Neither is there any way fo advantagious for draining moift ground, where there are no living Springs, as that of planting abundance of Timber, which will beft agree with that kind of foyl: for by this means, what was formerly noifome, and fuperfluous, is now converted partly into the ufeful aliment of the Timber, and partly fent abroad in infenfible exhalations, which (according to the nature of the emitting *Plants*) prove either very noifome, or wholfome to the Neighbour-Inhabitants. Great care therefore would be had in the choife of fuch Trees, as are to be planted in fuch moift ground, as are near to mens dwellings, or places of concurfe. They are not fools, who prefer *Firs*, and *Lime-trees* in their *Avenues* to *Oak* and *Elme*. Let the effects of the *Atomical* exhalations of *Alder* and *Oak* upon fine Linnen, and white Skins be more particularly noticed.

Having fpoken fomewhat of the aliment and growth of *Plants*, I fhall in the next place give a fhort hint at the motion of their aliment, efpecially of Trees. That the alimentary juice of *Plants*, is much thinner, than that of *Animals*, no man, I fuppofe, will deny, feing *that* is conveyed

veyed thorow the trunck, or body of the *Plants*, by imperceptible pores; but *this* (for the moſt part) is ſent thorow all the members, through patent and manifeſt Veſſels. But how the nouriſhing, and vital juice in *Plants* doth move, and by what paſſages, hath not yet been made known, by any that I have ſeen. I made once a few Obſervations, for trying of the motion of the aliment of Trees, which bred in me this conjecture. The nutritive juice of Trees is tranſmitted both to the roots and branches, through the heart, or pitch, and woody pores of the Timber, and when it is come to the extream parts, it returns again from the tops of the roots and branches, between the bark and timber, into theſe forenamed interior paſſages, and ſo back to the extremities again, and that continually, ſo long as the life remains. And becauſe the ſubſtance of that skin, or bark, which inveſts the fibres of the root, is more open and porous, than that which is upon the outward branches: therefore it ſeems, that ſo much as is ſuperadded to the ſtock of the former aliment, from the earth, is conveyed to the heart and pitch, by means of, and together with, that part of the retrograd juice, which returns from nouriſhing, and enlivening the timber of the root-branches, (for it is an eaſie Experiment, to make the top of any Tree become root, by laying it down) and receives the impreſſions of the life of the Tree, common to the whole maſs of alimentary juice, like the *Chyll* in *Animals* mixed with the blood of the *Veni-cave*, before it come to the heart.

This motion is not to be thought always alike ſwift, or of equal celerity: for the vital juice of the Tree becomes ſo thick and oleagenous in the Winter, that the motion thereof to the outward, is ſcarce diſcernable (though the preparation of the *Gemmes*, both for leaves and flowers,

are

are observed by the curious, and can be distinguished, even
in the coldest seasons) and the returns inward are in so
small quantities, that they are rather like vapours, than
liquid juice. Indeed, some Trees, when their root-branches are cut (even in Winter) will yeeld no small quantity of an acid liquor, which by addition of the recent *Lef-
fas* from the earth, smells evidently of the Matrix, from
which it did proceed. Moreover, the passages especially
from the branches to the Trunk, are so straitned and contracted, that the bark cleaveth to the Timber, as every
Wood-man knows. But so soon as the warm Spring hath
attenuated the ever-flowing juice in the whole Tree, then
doth it become turgid, and more aqueous over all: the
passages, and channels both in the trunk, and among the
tunicles, and particular skinnes, are so palpably filled with
this vital juice, that having no sufficient place to be com-
prehended in, it putteth forth new growths both in the
top, and in the root, which may be easily seen to have more
pitch than wood, and to be sealed on the extremity, with
the vestiges of a future Gemm ; that by the former, they
may the more freely receive the vital influences from within, and by the latter, may be secured from the depredation
of the external Air.

To prove the motion *ad extra*, or to the extremities
of the branches; take the branch of any ordinary Tree,
about the bignels of a mans wrist ; make it bare near the
body of the Tree of all bark, and subjacent tunicles (for
every Tree according to its kind, hath moe or fewer skins,
which serve for Veins, within the strong outmost Cortex)
at least for the breadth of a span, or two hand-breadth.
Then tye up the place, so excorticated with a *compost*, made
of horse-dung mixed with earth ; let it stand so from *May*,

till

till *November*. Then cut off the branch, a little above
the *Compoſt*, near the body of the Tree, and you ſhall
find it living and freſh, like the reſt of the branches: yea,
ſmall roots ſhall evidently appear to have come forth under
the *Compoſt* near the bark, but not under the bared place.
This branch in many kind of Trees being planted, will hold,
though not in all. I ſay then, ſeing the foreſaid bough is
nouriſhed from *May* till *November*, it is neceſſary, that it
receive nutriment from the body of the Tree, by the internal poroſities thereof: for the bark being diſcontinued
by *excortication*, can ſend nothing upward towards the top
of the bough; and if it received nothing from the root, it
would wither in a few dayes. Yea, leave the diſcovered
part naked, but for a few dayes, and of neceſſity the branch
dieth, the aliment thereof being exhauſted by the Air, before it can reach the extremities of the bough.

That the *Vital Balſome* of the Tree returns from the extremities by the internal bark, and inward ſuperfice of the
external, together with the ſmooth outward part of the
trunck, although the neceſſity of both timber and bark in
all *Inciſions*, and *Inoculations*, might perſwade the judicious, and the viſible courſe of the juice of the *Sycamor* in
February, and of the *Birch* in *March*, upon the cutting of
any ſmall branch, might convince any curious beholder;
yet the *knot* or *callus*, that is made upon grafted Trees, will
better inform the ignorant: for this *knot* being alwayes
upon the ſhoulder, or root of the Graff, and never upon
the top of the Stock, doth evince clearly, that it is made
by *reſtagnation*, of the deſcending, and not of the aſcending juice: otherwiſe, why doth it not ſwell the top of
the Stock, as well as the root of the Graff? Or why doth
it not extuberat in any other place of the Graff? Theſe

are

are accidental *varices*, which can hardly be shunned in Imping, seing the top of the Stock (except when it is very young and succulent) doth not receive so kindly, as it ought, the retrograd sap, although all that is sent out to the Graff must ascend thorow the pores of the Stock, Hence many times a considerable part of the Stock is mortified, because although abundance of aliment ascends to the head or top thereof, yet no more of it goes to the branches, but what is bestowed upon the Graff, a great part of the rest being exhaled by the Air (especially in big Stocks) and consequently, the place defrauded of its nourishment: no other wayes than when the motion of the vital sap faileth, either in the whole, or in part, a total decay or particular mortification of some part necessarily follows, as in the Stemms of annual Plants, and mortified tops of the *Ectrapelous* branches (that I may so call them) of *Willows*, *Plumbs*, &c. we may observe every *Autumn*.

OBSERVATION XXIV.

Sir,

" I Was not a little surprised, at the receit of yours, when
" I had considered your desire in it, being prest with two
" difficulties, which seemed equally hard to evite. The
" one, to give you my judgement in a matter wherein I
" have been so little conversant my self, and have had the
" steps of no other to follow, never one having hitherto
" touched that subject in writting ; I mean of *Coals*, and
" other *Minerals* of that nature, their *Course*, and other
" things relating thereunto ; the observation whereof (I
" grant) wants not its own pleasure, and usefulness. The
" other, to refuse the desire of a friend, when importuned,

" to

" to whom I owe my felf, by many obligations. This laſt
" having prevailed, hath determined me to affay the over-
" coming of the firſt. And though I am confident, what
" account I can give you , ſhall give but very little fatis-
" faction : yet I adventure to offer it, ſuch as it is, very
" freely in the following difcourfe , wherein you are not to
" expect, that I will meddle with fome queſtions , there-
" anent, which might be more curious, and pleafant, then
" profitable, or fatisfying, ſuch as, if *Coal*, and *Free-ſtone*,
" which keep one *courſe* , and have the fame accidental
" qualities , have been created in the beginning , in their
" perfection, as wee now find them , and fince that time
" only preferved , as they were created for the ufe of men,
" to whom all fublunary things were made fubfervient?
" Or , if they have been but produced gradually , as they
" ſpeak of *Gold* , and other *Minerals*, by the influence of
" the *Sun*, in the bowels of the Earth ? And if their pro-
" duction be of that nature , out of what matter they are
" formed ? Thefe things being above my reach, I ſhall leave
" their inquiry, to thofe that are knowing in the fecrets of
" Nature , and ſhall therefore give you a narration, of
" what either I have obferved of thefe things, which oc-
" curr in the *Winning* of *Coal* in my own experience, or by
" converfing with others of more experience than my felf,
" in doing whereof, I ſhall follow this Method.

First, I ſhall fpeak of thefe things that are common to all
Coal, wherein they all agree, and which are, as it were, *effen-
tial* to all, and of there differences, which are but accidental, and gradual fometimes, and yet are abundantly confpicuous, and caufeth different effects in the *working* ; as their
Dipps and *Riſe*, and *Streek*, for fo are they termed.

Secondly, of fome things, which are but accidental to *Coal*, and yet ſo ordinary, that ſcarcely any is found without them, in leſſer or greater degrees; ſuch are *Gae's*, and *Dykes*, that alter the natural *Courſe* of the *Metalls*, very incident to every *Coal*, though in ſome leſs frequent, conform to the nature and kind of the ground, where the *Coal* is.

Thirdly, I ſhall ſpeak ſomething of *Damps*, and of their different cauſes, and effects: of *Wild-fire*, and other ſuch like things, which are met with in the working of *Coal*.

And laſtly, of the beſt way for trying grounds to find *Coal*, where never any hath been hitherto diſcovered: of carrying on of *Levels*, for draining the water of *Coal* and making it workable.

It is to be coſidered, that all *Free-ſtone*, though of different natures, hath the ſame *courſe*, with the *Coal*, that ly either above them, or below them, except it be accidentally, interrupted: therefore, whatſoever is ſpoken of the one, is applicable to the other. And ſo we find in *Digging* or *Sinking*, that after the *Clay* is paſt, which keeps no courſe, all *Metals*, as *Stone*, and *Tilles* (which are *Seems* of black Stone, and participat much of the nature of *Coal*) ly one above another, and keep a regular *Courſe*; wherein the three things moſt remarkable are their *Dipp*, and *Riſe*, and their *Streek*, as it is termed.

The *Dipp*, and *Riſe*, are nothing but a declining of the whole body of the *Metalls*. And this general holds, that all of them from their *Center* riſes, till they be at the very ſurface of the *Earth*; ſome only at a foot or two foot, ſome at an ells diſtance from the ſurface, which is here termed a *Cropping*: and whether *Coal* or *Stone*, the nearer they come to the ſurface, the ſofter they become, till at laſt

laſt they are converted, if it be a *Stone*, to a very *Sand*, and if *Coal*, to a *Droſs*, which will not burn.

This declining or *Dipping*, of the *Coal*, is ſometimes greater, and ſenſible, ſometimes leſſer, and almoſt inſenſible. There being ſome, that if you conſider the declination, it will not be found one foot in ten ; ſome one foot in twenty, or one in thirty. Whereas in others it will be one foot in three, or one in five. And ſometimes it hath its *Courſe* from the *Center* of the *Earth*, almoſt in a perpendicular to the ſurface, it cutting it, near to a *right Angle*. The firſt ſort, they term *Flate-broad-coal*, in regard of the plainneſs, and evenneſs of its *Courſe*. The next, they call *Hinging-coal*. The laſt is called *Edge-Coal*. The firſt is the moſt profitable, in regard, that it's long before the *Coal-hewers* can reach the *Cropp*, and conſequently the more of it is workable. The ſecond and third ſort, are ſometimes of their own nature, more firm, and fitter for burning, but leſs of them can be reached in working. The *Courſe* of all the three is moſt perceptible in the three following Schematiſms.

Figure

Figure 1.

A   B

C

A   2   B

C

A   3   B

C   D

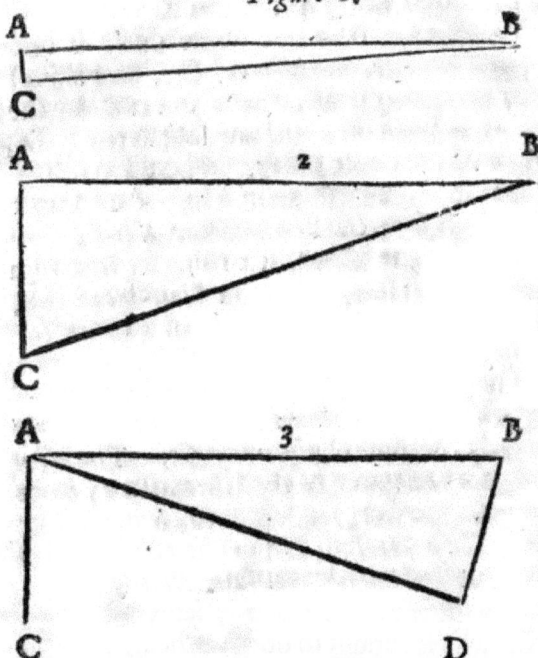

 In all the three Figures, the point B is the *Cropp* of the *Coal*. The Line B C is the body of the *Coal* declining or the *Dipp* from the *Cropp*. A C is the perpendicular, falling from the Horizontal Line, whereby the true declination or *Dipp* of the *Coal* is found. So that after you have found your *Coal* at B, you muſt ſet down your *Sink* at the point A. In the *Flat-broad-Coal*, which we ſuppoſe only to decline, three fathoms in ſixty; the *Sink*, that anſwers to the perpendicular A C, will be of deepneſs three fathoms.

thoms. If the diftance B A, be fuppofed to be 120 fathoms alongft the *Grafs*, or furface, then will the deepnefs of the *Sink* be fix fathom, and fo forth. In the fecond, if the *Coal* be fuppofed to decline one fathom in
three, the *Sink* A C, being fet down at the fame diftance
from the *Cropp* B, with the former, it will prove thirty fathom deep. If the faid diftance be doubled, it becomes
fixty fathom deep, and fo forth. In the third, keeping
that fame diftance alongft the furface, you fhall not encounter the *Coal* with a Perpendicular *Sink*, becaufe of its great
declination, and therefore through want of *Air*, and other
difficulties, you cannot dig fo deep, as is neceffary to that
effect, except the *Sink* fhould be made to decline, as doth
the *Line* A D. All thefe *Dipps* are to be feen in feveral
places of *Lothian*. The firft is moft confpicuous in the
Earl of *Wintons* ground at *Tranent*, where the *Coal*, and
other *Metals* are extraordinary flat and even. The fecond is within the faid Lordfhip of *Tranent*, in a piece of
ground, called *Wefter-Faufide*. The third in *Lonhead* of
Lafwaid, which pertains to Sir *John Nicolfon* of *Nicolfon*:
and in many other places, one may fee very different de-
clinations, who is curious to obferve them.

From this general pofition of the *Dipp*, and *Cropp* of all
free Metals, there is one confequent, which is no uncouth
Obfervation, namely that thefe *Metals* rifing from their
Dipp to a *Cropp*, every one of them rifeth in their proper
courfe, if none of thefe things whereof we fhall treat hereafter interveen, and make an alteration, that is the *Coal* or
Stone, which is loweft, comes farreft out in its *Cropping*,
which is eafily underftood by the fubfequent Schematifm.

Wherein

Figure 4.

A　　　D　　　F　　　H　　　K　　　M

C　　E　　　G　　　I　　　L

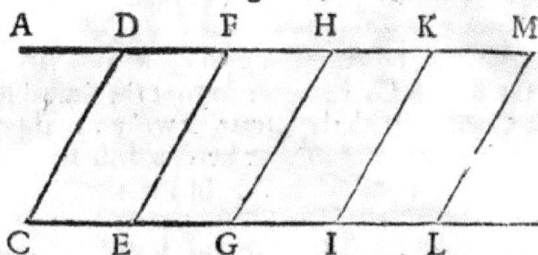

Wherein the Line A M reprefents the furface of the
Earth, C D, E F, G H, I K, L M, are fo many feveral *Metals,* lying in *courfe* one above another. Suppofe
C D were a *Stone,* and the *Roof* of the *Coal* E F (for fo
they term the *Stone,* immediatly next above the *Coal*) and
G H , I K , were other two *Stones* , interveening between the *Coals* E F, and L M, then if the *Cropp* of the
uppermoft *Coal* be found at F, the *Crop* of the *Stone* above
it, muft be found back, at the point D, and the *Cropp* of
the *Coal* under it, which is L M, muft be found at M. And
this diftance of *Cropp* is proportioned by the length of the
perpendicular between them, and the quantity of their declination. For, the more even and flat a *Coal* is in its
courfe, and other *Metals* , above and below , the farder
doth the *Cropp* of the loweft *Coal* advance before the *Cropp*
of the uppermoft. For illuftration whereof, let us fuppofe in two feveral grounds, two *Coals,* between which,
there is an equal diftance of perpendicular. And fuppofe
the *Metals* in the one ground to decline at 13 to 24, the
other at 13 to 16, then will the diftance between the
Cropps in the two grounds be very confiderable, as may be
reprefented by the two following Figures.

Figure

Figure 5.

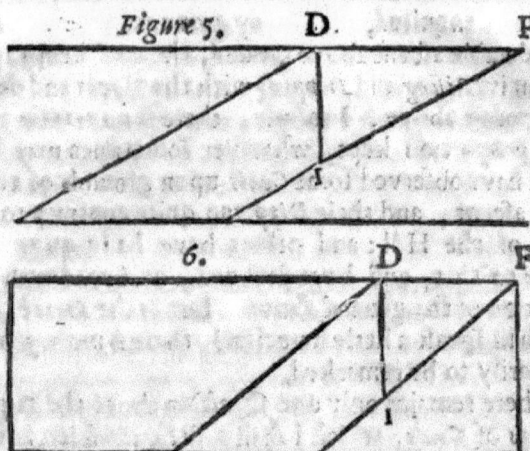

Suppose then, that D I, is of equal length in both *Tri-angles*, which is the perpendicular, between the two *Coals*: yet D F in the fifth Figure, is much longer than D F in the 6. And the reason is evident, because the *Angle* D I F, in the 5, is greater then the *Angle* D I F in the 6: and therefore the *Base* D F, which is *subtended* by the greater *Angle* in the 5, must be greater then the *Base* D F, which is *subtended* by the lesser *Angle* in the 6, which *Euclide* proves in his 24. *Proposition* of his first Book, and is demonstrat by *Proclus* in the *Scholium* to the 4th *Proposition* of the same Book.

By this is made to appear the profitableness of a *Flat-Coal*, beyond a *Hinging-Coal*, which was touched before, in regard that having the *Sinks* of equal deepness in both, there is much more of the *Flater-Coal* to be wrought, before it *Cropp* out, then of the *Hinging*, as there is a difference between the Lines D F in the first and second Figure, or between the Lines I F, in the same.

L l   If

If it be enquired, if in riſing grounds, where there is a conſiderable aſcent above ground, the *Coal* keeps a proportion in its *Riſing* and *Dipping* with the aſcent and deſcent of the ground above? I anſwer, there is no certain and conſtant proportion kept, whatever ſometimes may happen. For I have obſerved ſome *Coals* upon grounds of a conſiderable aſcent, and their *Dipp* run quite contrary to the deſcent of the Hill: and others have had a quite contrary *courſe* to that, and have declined, or *dipped* with the declination of the ground above. But in the *Streek* (whereof I ſhall ſpeak a little hereafter) there is more proportion ordinarily to be remarked.

There remains only one Queſtion about the *Dipps*, and *Riſings* of *Coals*, which I ſhall a little conſider, having encountered different judgements anent it, in converſing with perſons, who had experience in *Coal*, viz. whether *Coal* and other Metals, after they have declined ſuch a length from their *Cropp*, ſuppoſe from *Weſt* to *Eaſt*, take another *courſe*, and riſe to the ſame point, to which formerly they dipped?

Figure 7.

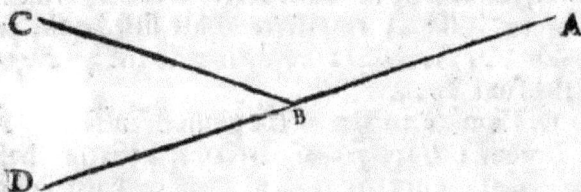

As if the *Coal dipped* from A, which is the *Cropp*, to B, which ſhould be the *Center* of that Body; and after that riſe to C? Or if it ſhould continue its declination thorow
B to

B to D, which is *Antipodes* to us ? I shall not offer to determine in a matter wherein there can be so little certainty attained, but shall give my opinion, which is founded upon the experience I have had, and Observations I have had occasion to make on that Head. And first, I find in all these *Coals*, wherein no contrary *Cropp* or *Rising* could be visible, there are invincible obstructions; as either, they have been near the *Sea*, and have *dipped* that way; and so if they took any contrary *course*, the *cropping* behoved to be in the *deeps*, and so no access to trace them. Or next, they have *dipped* towards the foot of a Mountain, and so the ground above rising the same way which they declined; their *course* could not be pursued, till a contrary rising should be discerned. Or thirdly, they have encountered some *Gae*, or *Dyke*, which hath cut them off, before they came to their full *dipp*, and thus their *course* was obstructed. Now, those that have been acquainted with no other *Coals* but such, I think it not strange, if it be hard to perswade them of those things they have not seen. But besides all those kinds, I have seen others, whose contrary *rising* and *dipping* have either been visible to the eye, or demonstrable by reason. For example, I have entered under ground, as it were at the point C, at the very *Grass-cropp*, and have gone following the *dipp* of that *Coal* to the point B, at which the *course* hath altered, and carried me out at the *Grass* at A, which are two contrary points of the Compass. And that alteration of *course* was not occasioned by any *Gae*, or *trouble*, which sometimes have that effect, the ground being very clean, and good *Metals*, keeping their *course* most regularly.

There are other instances for confirming my experience, in fields, which are so large, that 'tis impossible to work

the

the *Coal* so far to the *Dipp*, it falling deep, and so wants *Level* for conveying water from it, or wants Air, for following it to such a deepness, as to overtake its *Center*, where it takes a contrary *course*, and yet the contrary *Cropp* hath been wrought in several places, which is evident to be a part of the same body, with the other, both by the nature of the *Coal* it self, by the *Metals* lying above it, and the *Coals* below it, all which keeping the same *Course*, except when they encounter *troubles*, which are incident to some parcels of ground, more than to others. The greatest field I know wherein this is conspicuous, is in *Mid-lothian* where is to be found, the *cropping* of a *Coal* of a considerable thickness, which is termed their *great-seam*, or *Maincoal*, and the other *Coals* lying below it, which may be traced in the order following. At *Preston-Grange* these *Coals* are found *dipping* to the N W, and *rising* to the S E, which have been wrought up to *Wallifoord:* from that along by the foot of *Fauside* Hill, the *dipp* lying in the Lands of *Inneresk*, which marches therewith on the *North.* From thence it runs through the ground of *Carberry*, every one of these grounds from *Preston-Grange*, Giving *Levell* to another. From thence, through a part of the Lands of *Smeaton*, and next through a piece of ground belonging to the Family of *Buccleugh*, called *Coudon:* and through *West-houses*, which belongs to the *Earl* of *Lothian*, and at *Cockpen*, and *Stobhill*, from thence runs through to *Carington-Mill* ; all which is a *course*, which in *Streek* lyes near to S W, and N W, and will be in length about eight miles. From thence, the *course* of the *Coal* turns, and is found in the Barony of *Carington*, *White-hill Ramsay*, *Gilmerton*, and from thence taking its *Dipp*, quite contrary to what it had before, the other *Dipping* N and N W,

or

or N E, according to the turn of the *Streek*, it *Dipps* there S, S E, &c. and from *Gilmerton*, it is found at *Burntstone*, a piece of ground belonging to the *Earl* of *Lauderdale*: and from thence at the *Magdalen Pans*, where the turn of the *cropp* being within the Sea, is not seen, till it be found at *Preston-Grange*, where we began to remark its *course*. The parcel of ground, under which this great body of *Coal* lyes, is of a confiderable extent, it being eight miles in length, and five or fix in breadth ; in regard whereof many other *Coals* are found lying above the great *Coal*, the *cropps* whereof doth not come near the *Cropp* of it, by a confiderable diftance.

Though this inftance alone, may fufficiently convince, yet I fhall not be unwilling to give another. The parcel of ground, in which this *Coal* is found, is not of fo great an extent, as the other, and therefore its *course* may be the more eafily traced. For the greateft part, it belongs to the *Earl* of *Winton*, and lyes within the *Lordfhip of Tranent*, whofe contrary *Cropps*, are moft confpicuous. This great *Coal*, which is 10, or 12 foot thick (beginning at the head of the Toun of *Tranent*) where it hath been wrought, runs S W towards the march of the Lands of *Elphingfton*, belonging to the *Lord Regifter*, and continues in that fame *course*, till it come near to the *house*, and for the moft part *dipping* to the S E. And near the *house*, the *Cropp* is turned downward towards the march between *Elphingfton* and *Ormifton*, where the *dipp* is contrary to the former. And from *Elphingfton-mains*, it takes its *course* almoft round, through the Lands of *Panfton*, and returns to the Toun of *Tranent* where it began , which body of *Coal* will be in length two miles, and in fome places, as much in breadth. Now, I leave it to the judgement of any perfon, if there be

not

not more reafon to perfwade, that this fhould be the natural *courfe* of thefe *Minerals*, where fuch pregnable inftances, to evince it, are found ; then to conclude the contrary
from thefe *Coals* , the *courfe* whereof cannot be followed,
becaufe of the invincible impediments, I mentioned before.
However, I leave every one to be determined, by his own
opinion, and fhall be fatisfied to injoy my own, till thefe of
more experience convince me of the contrary.

There are fome other things farder to be remarked a-
bout the *Dipp*, and *Rife* of *Coals*, which (poffibly) every
one hath not feen, they being fo very rare , and therefore
are not fit here to be paffed without being confidered. One
is, of a *Coal*, which having that contrary *Dipp* and *Rife*,
(whereof I have been fpeaking) in one of the *cropps*, hath
not come out to the *Grafs*, and *terminat* ; but after it hath
rifen a confiderable way in its contrary *courfe*, in ftead of
Cropping out, hath taken a *Dipp* towards the fame point,
to which it *dipped* firft, and fo having *dipped* to the *Center*
of its *courfe*, it hath *rifen* again, and *cropped* to the contrary
point, as is to be feen in this eight Figure.

Figure 8.

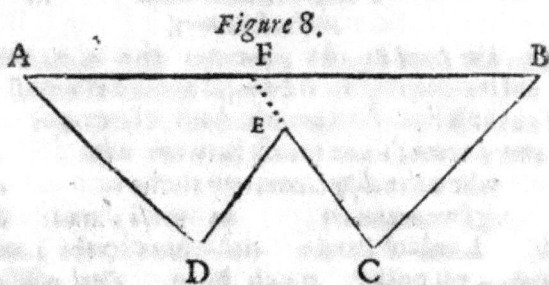

Where A B is the furface of the Earth. The point B
is the *Cropp* of a *Coal dipping* from N W, to the S E. From
C it

C it takes its *rise*, and *course* to a contrary **Cropp**, towards the point F, where the *dead Cropp* ought to be found. But in stead of going that length, it takes another *course* from the point E, *dipping* S E towards D, from which it takes its *rise*, and continues it to the point A, where it *terminats*, and where the *dead Cropp* is found. I grant, that it meets with a *trouble*, or *Gae*, at the point E, which seems to be the cause, why its natural *course* is changed. But its very extraordinary to see such an effect. But of this afterwards, in its own place.

There is yet another thing to be remarked, in the *dipps*, and *risings* of *Coals*, which is this. In the most part of *Coals*, that have their *course* from *dipp* to *cropp*, without the intervention of a *dyke or gae*, the declination is straight down, from the horizontal line drawn from the point of the *cropp*, to the fardest point of the *dipp*. That is, the *Coal* declining from that point in a right line, makes with the horizontal line, a *right lined angle*, *angulus rectilineus*, though in some the *angle* is more *acute*, and in others less, as is to be seen in the first, and second figures, where A B being the horizontal Line, and B the *cropp*, B C is the body of the *Coal* declining, which meeting with A B in the point B, constitutes a *right lined angle*, and where ABC in the second figure, is a greater *angle*, then A B C in the first. Yet I have seen a *Coal*, the body whereof from the *dipp*, or fardest point of declination, had its *rise* towards the *cropp* very insensibly, it being *Flatt*, and then began to be more sensible, till at last coming near to the surface of the Earth, it takes in a sudden such a *rise*, that from declining one foot of 12 or 14, it declines now one foot of three, as may be made evident from this following Figure.

Figure

Figure 9.

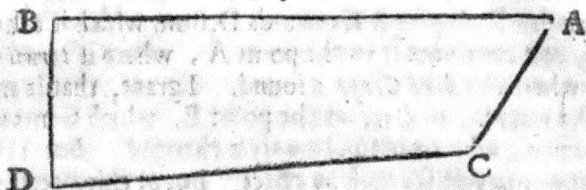

Where A B is the Line drawn from the extream points of the *Cropp*, right horizontal. The body of the *Coal* *rising* infensibly, is D C. But affoon as it comes to C, it rifeth with a great afcent till it *Cropp* out at A. Here you fee, that in ftead of one fide of a *Triangle*, which the *courfe* of other *Coals* in their *rifing*, or in their declination makes, this *Coal* in *rifing* makes two fides, namely D C, and C A, the Figure D B C A being quadrilateral. The *Coal* of this *courfe* was really wrought, and is yet vifible in its *wafte*, where there is found no *Gae* or *Dyke* to make this alteration.

Thefe are the chief things that I have thought worthy of Obfervation in the *Dipps*, and *Rifings* of *Coals*, and therefore I come now to touch a little the other part of their *courfe*, which is commonly termed the *Streek* of a *Coal*. To make intelligible to thofe, who are not experimentally acquainted with *Coal*, this term, or what the *Streek* is, we muft lay this foundation, that the *Coal* is a *Phyfical Body*, and fo hath its three principal dimenfions, which do conftitute it fo, *viz.*, *Longitude*, *Latitude*, and *Profundity*. Its *Latitude*, is that part contained between its extream lines, which is meafurable by its furface, to which

its

its *dipping* and *rising*, though alwayes incident, yet is but accidental. Its *Profundity* is to be meafured by the diftance, between the two furfaces, immediatly next to it, above and below: which are termed in *Coallery* its *Roof* and *Pavement*, becaufe of the refemblance they have to the *Roof*, and *Pavement* of a houfe. The *Longitude* is nothing elfe but what is termed by the *Coal-hewers*, the *Streek*. For if you imagine a Line drawn along the extream points of the *Rife*, or *Cropp* of the *Coal*, that is properly the *Streek* of the *Coal*.

There are but few things to be remarked, as to this part of *Coal*: only firft to find how it lyes, to what points of the Compafs it moves. For knowing whereof, there is this general Rule, that, having found your *Dipp* and *Rife*, to what ever Points that *Courfe* is directed, the *Streek* is to the quite contrary. For fuppofing a *Coal Dipp* S E, the two points, that refpect the *Dipp* and *Rife*, muft be S E, and N W, being the points oppofite one to another. Then it muft needs follow, that the *Streek* muft run S W, and N E, which two *courfes* divides the Compafs, at *right Angles*. And therefore, where a *Coal* is found to have contrary *Dipps*, and *Rifings*, they declining fometimes to all the Points of the Compafs (whereof there hath been given two notable inftances before) it muft needs follow, that there be alfo contrary *Streeks*, and fo the *Streek* of a Body of *Coal* is fometimes found to defcribe a round figure, though not perfectly circular, and fomtimes a *multangular* figure. For it cannot be fuppofed that the *Streek* makes alwayes a right Line, between the two points, from which it is reckoned. For example, between the *Laird* of *Prefton-grange* his houfe at *Prefton-pans*, and the *Stob-hill*, there are the *Streeks* of feveral *Coals*, lying one a-

M m

bove

bove another, which will be of length, about seven or eight miles, lying near upon S W, and N E ; yet the *Cropps* of the said *Coals* (their *dipp*, and *rise*, being N W, and S E) are sometimes farder advanced towards the S E, sometimes farder back towards the N W, by the difference of a mile, and this generally occasioned by the encounter of a *Dyke* or *Gae*, whereof hereafter.

The same question, that occurred in the *Coals dipping* towards a Hill, or *rising* above ground, comes to be inquired into here ; *viz.* If a *Coal* encountering an ascent, or *Brae* above ground in its *Streek*, rises also with the ground, and keeps its ascent? I answer, I have found it so in all the *Coals* I have ever seen of that nature, GOD in his providence, having so ordered it, that thereby it may be the more useful, in regard more thereof may be wrought by one *Level* or *Aquæduct*, by which the Water is conveyed away, as afterwards will be observed in speaking to *Levels*. For confirmation whereof, I shall bring instances both of *Coals*, that declines towards the Hill, and of others that declines with the same *dipp*, the Hill hath it self. In the *Coals* of *Bonhard*, *Grange*, *Kinglassy*, and *Kinneil*, which keep all one general *course*, the ascent above ground is from the Sea, (which lyes North) towards the South, or thereabout ; the *Coal dipps* or declines towards the N W, and so consequently *rises* to the S E. The *Streek* of these *Coals*, is from the N E to S W, which slops alongs the Hill, and comes up to the top thereof to the Westward of the House of *Bonhard*. Now, in sinking in that ground, if an equal proportion be kept, in all the *Sinks* from the *Cropp*, and a just allowance given for the different *Rising* above ground, the *Sinks* will be near of an equal deepness along all the *Streek*. So that a *Sink* upon the
same

same Coal near to the Sea, which is the N E point of the
Streek, at equal distance from the Cropp, will be as deep as
a Sink upon the top of the Hill, being the S W point of
the Streek at the same distance from the Cropp, allowing
alwayes the different rise above ground, and excepting
some particular troubles falling in upon the Metals of one
Sink, and not of another, and so making them dipp more,
which will occasion a difference of the deepness. The same
is also found in the Coals of Dysart, and Weems. As also
in that great body of Coal before mentioned, between
Preston-grange and Stobhill, the declination whereof is to
the N E, which is also the course of the descent above
ground.

Another instance is from the Coals within the Lordship
of Tranent, the dipp whereof is of another course, being
contrary to the descent of the Hill, viz. the Coal dipping
to the S E, and consequently the Streek running S W,
and N E, where the same is to be observed that was seen
in the other, anent the equality of the deepness of Sinks
along the Streek, with the same allowances, and exceptions
before mentioned.

Some have been of opinion that Streeks of Coals ly generally South and North, or to some of the points near to
these two Cardinal ones, between South and S W, and
North and N E, as South and by West, and North and
by East, &c. To which general I cannot agree, in regard of what I have before made evidently appear, viz.
that some Coals have their croppings towards all the points
of the Compass, and the Streeks being regulated by the
Cropps, they must necessarily be judged to have their courses
proportioned to theirs: so that if a Coal dipp to the true
North, and rise to the South, the Streek must be East, and

West.

Weft. However, I acknowledge two things, for confirming that opinion.

Firft, that of all the Coals I ever have feen, where thefe
contrary dipps and rifings, could not be traced, and made
vifible, theSreek hath inclined to thofe points of South and
North. But I muft alfo confefs, that they are but few I
have feen, in refpect of what I have not feen, and fo if any
others experience, who have feen more, contradict mine, I
fhall willingly yeeld, and not be tenacious.

Next, in thefe Coals, which I inftanced, that have their
Cropp to all the Points, and confequently their Streeks, and
in others of the fame nature, which I have feen, and not inftanced, I found that part of the Streek, which lyes towards
thefe Cardinal points, to be the greateft, being double, or
triple to the other Sreeks in length. So that when the
Streek, that lyes either along the one Cropp, or the other,
towards the S W, and N E, will be feven miles in length,
that lying S E, and N W, will be but four, and fometimes
lefs. And this is all the account I can give, of that part of
Coal, called the Streek.

The fecond thing I promifed to fpeak of, was of fome
things, which are but accidental to Coals, and yet fo ordinary, that hardly are any found without them in leffer,
or greater degree, fuch are Gae's, and Dykes, which alters
their natural courfe, and they being the occafion of fo much
Trouble, in the working of Coal, and following its courfe,
the Coal-hewers call them ordinarily by that name Trouble.
This Trouble or Gae then, is a Body of Metal falling in upon
the courfe of the Coal, or Free-ftone, obftructing, or al-
tering their kindly and natural courfe, keeping no regular
courfe it felf, and being of nature alwayes different from the
Metal, whofe courfe it interrupts. And thefe Gae's differ

fer also among themselves, in their nature, and in their
course they keep: or more properly in the way wherein
they encounter other *Metalls*, and in their effects. In
their nature, for some of them consists of an impregnable
Whin-Rock, or *Flinty-Stone*, thorow which it is almost
impossible to work: and if there be a necessity to cut them
thorow, it is done at a vast expence, and takes a long time,
and must be cut open to the surface of the earth, it being
impossible to Mine it under ground. Some of them are
again of *Stone*, like a *Free-stone*, but seems rather an abor-
tive of nature, they having no rule in their *course*, by which
a man can follow them, nor can their stone be useful.

In their encountering of *Coals*, or *Free-stone*, sometimes they encounter them in the *Dip*, and sometimes in
the *Streek*, and sometimes between the two. These that
are met with in following the *Dipp* of the Coal, ly along
the *Streek* thereof. For example, if the *Coal Dipp* S E,
the *Gae* lies N E, and S W. These that are encountered
in the *Streek*, lyes to the *Dipp* and *Rise*: so the *Coal Streek-*
ing N E, and S W, the *Gae* is found to ly S E, and N W.
Others of them, lyes between *Streek* and *Dipp*, that is to
some point between the two: as the *Streek* being S W,
and N E, and the *Dipp* and *Rise* S E, and N W, there
may be a *Gae* found lying W S W, and E N E. Now,
when I speak of a *Gae's* lying to such Points of the Compass, this doth not contradict what was said before, that
they had no regular *course* themselves. My meaning being, that though they have a certain length, lying between
two points, and a thickness between two *Metalls*, yet by
the *Metal* of the *Gae* it self, it is impossible to know its
course, as it is in other *Metalls* of *Coal* or *Free-stone*, whose
courses are discernable at the first view,

Their

Their effects are different, as their nature and *courſe* are different : only they agree in theſe two generals. Firſt, that all of them renders that part of the *Coal*, that comes neareſt to them, unprofitable and uſeleſs, though ſome leſs, and ſome more, they being unfit for burning. And it is remarked, that theſe *Gaes* that conſiſts of *Whin-rock*, renders the *Coal* next to it, as if it were already burnt, being ſo dried, that it moulders in handling it. In others, the *Coal* is not altogether ſo ill, and yet its nature is altered, from what it is at a diſtance from the *Gae*. The next general is, that all of them alters the natural *courſe* of the *Coal* in leſs or more, ſome of them making it *Dipp* much more then its ordinary *courſe*, which they call *Down-gaes* : Some again making their *riſe* much more than their *courſe*, which they call *Up-gaes*. Others making an alteration as to the *Streek*, cauſing it go out beyond its ordinary bounds, as we obſerved before in that great *Streek* of *Coal* between *Preſton-Grange* and *Stobhill*.

Now it is to be conſidered, that when in working of a *Coal*, whether to the *Dipp*, or *Riſe*, or *Streek*, one of theſe *Gaes* is encountered with, the *Coal* is quite cut off, and as it were *terminat* : ſo that you ſee nothing where the *Coal* ſhould be, but either a *Stone*, or *Clay*, or rotten *Till*, or ſome ſuch thing. And the practique of *Coallery* is to trace the *courſe* of the *Coal* through that, till you overtake it in the other ſide. And before any thing be ſaid to that part, you muſt notice, that ſome *Gaes* are of greater force than others, and their influence upon the *courſe* of other Metalls greater, whence you ſhall ſee a threefold effect. One is, that by ſome great *Gaes*, which a *Coal* meets with, it is quite cut off, ſo that in the other ſide thereof, there is not a veſtige of that *Coal*, or of any other Metal that was

above

above it, or below it, to be seen. And if there be any other *Coal*, as sometimes there are, they are quite different from them of the other side. I said by *some*, because there is one instance to the contrary, which is somewhat singular. In the Earl of *Winton*'s ground at *Cockeny*, there is found a *course* of Coals and *Free-stone*, *dipping* to the S E in the *Links*; and upon the *full-sea-mark*, there is a *tract* or *course* of *Whin-rocks* lying E and W, underneath which these Coals and *Stones* comes thorow without alteration of *course*, and are found within the *Sea-mark*, with the same *Dipp* and *Rise* upon the North side, they had upon the South side of the said *Rocks*: and yet the *Coal* is encountered upon the South hand by a *Gae* under ground, through which it passeth, not without a considerable alteration.

The greatest of these *Gaes*, that I know, is that which takes its beginning, that we see on Land, at the *Harbour* of the *Pans*, called *Achisons-Haven*, which hath been cut by *Preston-Grange*, for *Level* to his *Coal*, and goes from that to *Seton*, which may be traced above ground, almost the whole way; and hath been cut at *Seton*, for serving the *Level* of that *Coal* now wrought at *Tranent*. From thence it passeth through the fields of *Long-Niddry*, a place pertaining to the Earl of *Winton*, and through the *Coats*, which pertains to the Earl of *Hadington*, till it joyn with *Pancreck-hills*, a tract of Rocky Mountains, from whence it is traceable to *Linton-bridges*, where it is visible in the Water, the Water of *Tyn* falling over it, and making a *Lin*, which they call *Linton-Lin*; from thence to the *East-sea*. And it is known by Sea-men, that it keeps a *course* thorow the *Firth* from *Achisons-haven*, (whence we reckoned its beginning upon Land) towards the West

and

and N W, it being found to the Southward of *Inch-keith*, and before *Leith*, where ſtands a *Beacon*, and ſo can be traced to the North Shore.

The ſecond effect of *Gaes*, is to cut off the *Coal* quite, as to a part of the field, ſo that in the other ſide, having pierced the *Gae*, you ſhall not find the *Coal*, and poſſibly not within a quarter of a mile of the *Gae*, which cuts it off, and at that place ſhall only find the *Cropp* and the Body *Dipping*, as it did before it was cut off; and if you ſhall meaſure between that ſide of the *Gae*, where you loſt your *Coal* (I ſuppoſe the *Coal* then being 24 fathom from the *Graſs*) to the place where the *Coal* in the other ſide of the *Gae* ſhall be found at the ſame deepneſs, it will be near 500 paces. For making this more intelligible, let us ſuppoſe a *Coal Dipping* S E, and in working to the *Dipp*, there is a *Gae* encountered with (This was really done in a piece of ground I know, and ſo it is no meer ſuppoſition) at which *Gae* the *Coal* is cut off; for finding whereof the *Gae* is pierced, and nothing found in the other ſide, *viz.* in the S E ſide of the *Gae*, but at more than 100 paces diſtant, the *Crop* of a *Coal*, which lyes under the *Coal*, that was loſt, was found, after which it was eaſie to find the other. Now, that it was the ſame *Coal*, that was loſt, upon the North ſide of the *Gae*, is not only evident, by the kind of *Coal*, and all the *Metals* above, and below keeping the ſame *courſe*, but by this, that the *Gae* wearing out towards the Weſt, the two parts of the *Coal* that was ſeparated by it, joynes themſelves again, and continues in one body, as they were before ſeparation.

The laſt effect of the *Gae* is, that it doth not quite cut off the *Coal* from the other ſide of it, but makes an alteration in the *courſe*, either in the *Dipp*, or in the *Riſe*, or

Streck,

Streek, as was before noted : so that in meeting with one of these *Gaes,* having considered its nature, and pierced it, the Coal will be found in the other side, immediatly touching the *Gae,* but with an alteration of *course.* Now, in these two last effects, since the Coal is not totally cut off, it will be worth the inquiry, to find the surest way of recovering the Coal after it is lost. Therefore, where the Coal is not cut off, by a considerable distance, and having pierced the *Gae,* it is not to be found in the other side, you are to consider well the nature of the *Metals* you find approach to the *Gae,* and if they be such, whether *Stone,* or *Coal,* as you know to ly under the Coal that you have lost, then you may be sure the Coal is to be found above in its *course,* which is to be traced by the *Dipp* of the *Metals* you find. As sometimes I have seen, when a Coal hath been cut off by a *Gae,* happly there is another Coal under it 12 fathom, after the *Gae* hath been pierced, and the lost Coal not coming near to it in the other side, that hath been found there, by which it was certainly concluded, that the uppermost Coal behoved to be there also, though a little back, conform to its *course.* But, if the *Metals* or *Coals,* under the lost Coal, hath not been known, then you are to take notice of the *Dipp* and *Rise* of these *Metals,* you find on the other side of the *Gae,* which you have pierced, and making that your rule, *range back over the Metals,* conform to the direction to be given afterwards, and you shall find the *Cropp* of the Coal you want, and after which you were inquiring.

Where the Coal is not quite cut off by the *Gae,* but hath its *course* only altered, you are to consider, in searching for it, before you pierce your *Gae,* that which the *Coal-hewers* term the *Vise,* or some of them the *Weyse* of the *Gae,*

N n   which

which in effect is nothing else, but a dark vestige of the
Dipp or *Rise*, that the body which now constitutes the
Gae, should have had naturally, if it had been perfected;
which when it tends downward, then must the *Gae* be put
over that way, and in the other side shall the *Coal* be found,
and *Down*, as they term it; that is, the *Dipp* which it had
naturally, augmented. And, if the *Rise* be *Up*, the same
way must be taken for piercing the *Gae*, and the *Coal* will
be found *Up*, that is, its *Rise* augmented. But these things
cannot be made so intelligible, as by seeing, there being
many things in the alteration of the *course* of Metals very
curious, and worthy of Observation : as when a *Coal* is
cast down out of its natural *course* by a *Gae*, and so made
sometimes *under-Level*, it riseth as much to another hand,
and the *Cropps* go so much farder out, which still makes the
Level useful, the use whereof would have been judged lost
by the *down-casting*. Sometimes a *Coal* made to have four
contrary *courses*, as is evident from the eighth Figure,
where there being a *Gae* at E, makes it take such another
course, in stead of coming out to the *grass*. Sometimes,
before the Metals overtake the *Gae*, they are made to ly
like a *Bowe*; one instance whereof is visible above ground
in some Metals lying between *Bruntiland* and *Kinghorn*, at
a place called the *Miln-stone*, where there is a small *Coal*
with *Free-stone* above it, all *Dipping* to the S E, and *Rising* to the N W. Upon the *Rise* they meet with a *gae*,
which is a great *Whin-rock*. In their *course* to the *grass*,
before they touch the said *Rock*, they take a contrary
course, and *dipps* into it, and are there quite cut off. The
manner whereof is to be seen in this tenth Figure following.

Figure

Figure 10.

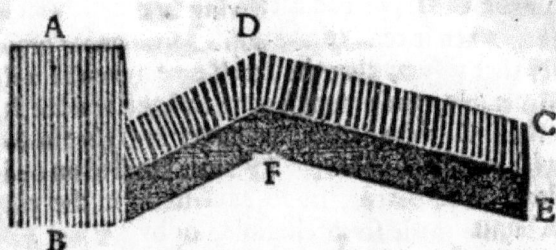

Where A B is the *Rock*: E F the *Coal*: C D the *Free-stone*. Now, whereas they ſhould have riſen towards A, they turn at D, and *dipps* into the *Rock*, which any may obſerve in paſſing that way. Many other ſuch motions are obſervable, which I paſs, and leaves them to the obſervation of the curious.

The third thing I promiſed to ſpeak of, was of *Damps*, and as they are termed by the *Coal-hewers*, *Ill Air*. Theſe do deſerve a more accurat inquiry into their *kinds*, their *cauſes*, and *effects*; then I am capable to make, there being many things in them very conſiderable, and worthy of a narrow ſearch: therefore following the courſe I have hitherto obſerved, I ſhall ſhew my own Obſervations thereof, and leave the more curious ſearch to the ſpirits fitted for that purpoſe.

This *Damp* then makes an obſtruction of reſpiration in Men, or other living Creatures, in Subterraneous ſpaces, as *Caves*, *Coal-rooms*, *Levels*, *Sinks*, and ſuch like; which obſtruction proceeds principally from two cauſes, both which goes under the name of *Ill Air*, among the vulgar. The firſt is the corruption, or putrefaction of the Air, whereof there are two ſorts; one is in places where

hath

hath been *fire* kindled, which burns the *Coal* under ground, the fmoke whereof, being full of Sulphur, and other Bituminous matter, and not having free paffage to come above ground, filleth all the *wafte Rooms* under ground, and infects the Air fo, that the fmell of it, even at a diftance, is intolerable, and amongft it no living Creature is able to breath. Of this there are examples in *Dyfert* in *Fife*, and *Faufide* in *Eaft-Lothian*. This was kindled on defign by a Fellow, who for his pains was hanged in the place, and hath burnt thefe 50 years, and more, the *fire* whereof is fometimes feen near the *grafs*, with abundance of fmoke, as it runs from one place to another. The fecond, where the Air is corrupted without the mixture of fmoke, or any other grofs corrupting body, which is the moft confiderable of all *Damps*, and hath the ftrangeft effects, in killing *Animals* in an inftant, and fo hath been alwayes moft prejudicial in the works, where it is found, many perfons having thereby loft their lives, without accefs to cry but once *Gods mercy*, to fome inftances whereof I have been witnefs. I fhall not offer to determine about the caufe of this *Damp*, but fhall give an account of fomethings I have obferved about it, which when duely pondered, may haply lay a foundation, at leaft of a probable conjecture, whence it may proceed.

This kind of *Damp* then, and *Ill Air*, is never found in *Coal*, or other *Metals*, where there is Water to be found; I mean, whence the Water hath not been drawn away by a *Level*, or *Aqua-duct*: as in *Coals*, where there is a necefity to *lave* the Water from place to place, or to pump it along the afcent or rife of the *Coal*, to the bottom of the *Sink*, from which it is drawn out above ground, this *Ill Air* is not found. Nor is found frequently, if at all, in
thefe

thefe *Coals* where the Water is drawn from the *Coal* by
a *Level* or *Aqua-duct* under ground, till it come of its own
accord to the bottom of a *Sink*, which is in place of a *Ci-*
ftern, out of which it is forced alfo above ground, and dif-
fers only from the other, that the Water runs here of
its own accord by a defcent to the *Sink*, which is termed a
drawing Sink: in the other it muft be forced by the *Rife* of
the *Coal*, becaufe happly, a *Sink* upon the *Dipp* would be
of fuch a *deepnefs*, that no force could draw it up in a perpendicular.

But this kind of *Damp* is found ordinarily in thefe *Coals*
from which the Water is drawn by a *Level*, the beginning or mouth whereof is above ground, and carried along
by a right Line under ground, till it overtake the *Coal*,
which it is to dry: fo that the Water which comes from
the *Coal*, runs without being forced, and is fometimes fo
confiderable, that it makes *Mills* go, without any other
addition, as is to be feen in the Earl of *Winton's* Lands of
Seton, where four *Mills* goes with the Water that comes
from under ground, out of the *Coal*; which kind of *Le-*
vels are only found where the *Coal* lyes in a Field, which
hath a confiderable *Rife*, or afcent above ground ; there
being a neceffity to make ufe of the other two wayes fpoken of, for drying the *Coal*, when the Field in which it
lyes is a Plain.

Further, of thefe *Coals*, which are dryed by the *Free-*
level (for fo they term the *Level* that runs unforced)
there are fome to which this kind of *Damp* is more incident,
than to others. The caufe of which difference is found to
be, the folidity and clofsnefs of the *Metals*, whether of
Coal or *Stone*, wherein fome exceeds another. There being fome, that are full of *rifts*, or empty fpaces (I mean
empty

empty of any part of the ſame body where they are) which
will ſometimes ſerve, to convey a conſiderable quantity of
Water in place of an *aqua-duct* or *level* ; which ſpaces are
termed by the vulgar, *Cutters*, which ſometimes proves
very profitable in the ground where they are found, both
in regard of the uſe they ſerve for, in ſtead of *Level*, and for
rendring the *Metals* wherein they are found, more eaſie to
work, in making them yeeld eaſily to the force of the
wedge and *leaver*.   Other Metals there are, wherein few
of theſe *Cutters* are to be found, and if water be to be conveyed through them, there is a neceſſity of cutting a paſſage through them for that effect.   Now, this *Damp,*
whereof we ſpeak is found moſt frequently, and moſt violent in the fiſſt ſort of *Metals*, *viz.* in theſe which are
full of *Cutters* or *Rifts*, which gives ſome ground to this
conjecture of its cauſe.   Theſe *Spaces* which are found in
Coal, or other *Metals*, as *Stone* or *Till*, before the *Coal*
begin to be dryed by a *Level*, are full of water, which is
ſtill in motion, as are all *ſubterraneous ſprings*, whereof
ſome are more violent, ſome more ſlow, conform to the
paſſage they have to the fountains above ground, where
they diſcharge themſelves.   Now, for drying theſe *Coals,*
and rendring them workable, there is a neceſſity to cut a
paſſage, thorow which that water diſcharges it ſelf quickly, it being large, and admitting a great quantity at once,
by vertue whereof ; a great field is drained at once, and
the *Sourſe* not being able to furniſh ſo much water, as the
Conduit is able to convey, theſe *Spaces* in the body of the
Metals, being emptied of Water, muſt needs be filled with
Air, which *Air* having little *contact* and *commerce*, with
the great body of *Air* above ground, and ſo hath little or
no motion, corrupts in theſe places, and thereby becomes
poiſonable

poisonable, so that when any *Animal* is necessitat to draw
it, and *respire* by it, it choaks them on a sudden, just as
standing Water, which being without motion corrupts,
and becomes poisonable, though haply not in so great a
degree as the *Air*: the *Air*, being a body much finer and
purer, than Water, that holding good in it, *corruptio
optimi pessima.* This is much confirmed by what is before
afferted, that in the *Coals*, whence the Water is drawn, and
they drained, but not by *free-course*, but by *Force*, as
Pumping, and drawing by *buckets*, these *Damps* are seldom
or never found: because the passage of the Water being
forced, it does not so suddenly dry the *Metals*, as the other,
whereby there is alwayes left in these *Spaces* some Water,
which being it self in motion, keeps the *Air* also in motion with it, and thereby the *Air* is kept from *corruption*,
at least in such a degree, as it is in the other. Hence we
find, that in these kinds of *Coals*, the *Rooms* under-ground
are alwayes wet, or for the most part they are so: whereas
in the other, there will be no Water found to wash a mans
hands: and sometimes the *Coal* through want of Water,
becomes so dry, that it cannot be wrought in great pieces,
as others, but crushes in the very working, and when
wrought, is rendered useless, and will not at all burn.
This puts me in mind of a very pleasant conception of a
worthy and learned Person, Doctor *George Hepburn* of
Monk-ridge, with whom I had occasion one day to discourse
on this Subject. He is of opinion that the *Water* is the
Mother of the *Coal*, whereby it is preserved fresh, and in-
corrupted, and that when the *Water* is drawn off, and this
Damp follows, it is not the *Air*, which succeeds in place
of the *Water*, and is corrupted for want of motion, that occasions it. But as we see, when the corruption of a Li-

quor within a Veſſel, when the *Mother* is gone, corrupts the Veſſel it ſelf, and occaſions an ill ſavour or taſte in the Veſſel ; ſo that the *Coal* being corrupted by the want of its *Mother*, the *Water* ; corrupts the Air in the ſubterraneous *Spaces*, as in *Coal-Mines, Sinks, Caves*, and other ſuch like. He had likewiſe another pleaſant conception about the *generation of Coal*, judging it to be formed gradually out of another Metal, as of *Till*, by the help of *Water*, of which he himſelf may perhaps give an account. And though I be not of his opinion in that matter, yet I muſt acknowledge, I was taken with it, and ſhall be glad to ſee a more full account of it from him, than he had acceſs to do in the ſhort conference we had.

The effects of this *Damp* are firſt, it hinders the burning of all combuſtible matter, as *Candle, Coal, Pitch, Sulphur*, &c. ſo that if you take a *Torch* lighted, and let it down to a *Sink*, where the *Ill Air* is prevalent in the time, it ſhall ſtraightway extinguiſh it. Or take a *Coal*, which is burning, and let it down, it ſhall not only extinguiſh the *Flame*, but ſhall make the *Coal* in an inſtant *dead*, and as cold as never heat had been in it. But the moſt dangerous effect is, its killing of *living Creatures*, whereby many perſons have been ſuddenly killed. Some in going down to a *Sink*, where it hath been powerful, have fallen out of the *Rope*, and periſhed. Others have been choaked, and yet have gotten out by the help of others in a ſudden, and have remained a conſiderable time without the leaſt appearance of life, but yet have at laſt recovered. Yet it hath been obſerved, that ſome of theſe perſons that have been ſo ſtruck with the *Damp*, and recovered, have had alwayes ſome *lightneſs of Brain* thereafter, and never ſo ſettled as formerly. This I know to have happened to one, whom I have ſeen ſo, many times thereafter.   What

What hath been its effects on some *Animals*, whereof you have made Experiment, I leave to the account you have given. One thing I shall only mention, which to me seems somewhat strange, that notwithstanding these *Damps* are so effectual, and causeth so suddenly the death of *Animals*, yet the *Ratts*, which are in some of these places, where the *Damps* are most violent, are not reached by them. For sometimes, when they are so powerful, that nothing that lives can enter under ground, without sudden death, yet they continue there, and are not found to diminish, even where they have no access to escape, by coming above ground. Or if it should be imagined, they removed to some other place of the ground, where the *Damp* is not, how is it, they are not as quickly choaked with it, as *Dogs* are, and other *Animals*, which at the first encounter are killed?

If it be inquired, how comes it to pass, that in these Fields of *Coals*, which are dryed fully (as was said) and to which these *Damps* are incident, because of corrupted Air that remains within the Body of the *Coal*, or other *Metals*, how comes it to pass (I say) that they are but sometimes incident, and are not alwayes found? For clearing this, it is certain, that even in the grounds, where these *Damps* are most frequent, for the reasons above mentioned, yet they are only powerful when the Wind blows from such a certain *Point*, as some Chimneys, that do only smoke, when the Wind is in such an *Airth*. This is so generally, and well known, that the *Work men* observe it, and when they find the Wind in such a Point, whence they fear the *Damp*, they will not enter under ground, till tryal be made of the Air, which they do in *Sinks*, by first letting

ting

ting down a lighted *Candle*, or some burning *Coals*: which if they do not burn, then there is no access to enter.

Secondly, the wind in which this *Ill Air* is most noxious, and hurtful, blows from that Point, where the *Field* of *Coal* lyes, that's not yet wrought, which seems somewhat strange, and yet when duely considered, it will appear abundantly consonant to reason. An example of this is to be found in the *Coal* of *Tranent* and *Elphingston*, the *Streek* whereof goes to the *rise* of the Hill above ground, from N E to S W, as hath been formerly observed. So that the beginning of their *Level*, is at the N E point of the *Streek*, from which the *Coal* hath been wrought up along the *Streek* towards the S W, the *Wastes* lying all towards the N E. Yet when the Wind blows from N E, or N, or almost from any other Point of the Compass, they are not troubled with this *Damp*. But if it blow from S W, and blow hard, they are in hazard to encounter it. And though the *Damp* is not alwayes found when that Wind blows (whereof there may be some particular cause) yet it is never observed in another Wind, whether it blow less or more: the reason whereof may probably be, that the Wind blowing from other Points, as from N, or N E, hath more access to enter the *Wastes* under ground, and move the Air that is in them, towards the face of the unwrought *Coal*, whence is supposed to proceed the corrupted Air, that lurks in the *Rifts* and *Cutters* thereof, (from which the Water is drawn away,) and occasions the *Damp*. Now this Air being moved by the force of the Wind, keeps the corrupt Air from coming out, it being stronger then the other. Whereas, upon the contrary, while the Wind blows from S W, it entering the empty *Rooms*, drives the Air under ground from the face of the unwrought

wrought *Coal*, down towards the *old wastes*, which have their *course* from the beginning of the *Level*. By which means, the Air, that is corrupted within the bowels (to speak so) of the *Coal*, comes out to the *Wastes*, without resistance, it being certain, that *Fluid Bodies*, as *Water*, and *Air*, inclines to move towards that place, where they meet with the least resistance. Hence is it, that the more direct the Wind be, in blowing against the face of the un-wrought *Coal*, as is the Wind from N E, the *Ill Air* is the more repelled and driven back, but the more oblique it be, as are the Winds from these Points, that are nearest to S W, the *Air* is not so good and free: which difference is known by the burning of *Candles*, they burning with greater difficulty in these Winds, than in others, which blow from these Points nearest to N, and N E. Some are of opinion, this *Ill Air* (in those places we have been speaking of) comes from the great *Wastes*, that ly above the un-wrought *Coal*, and by strong S W Winds is driven thorow the *Cutters* thereof. Or the Wind blowing from that Point, and coming thorow these *Cutters*, brings the corrupted Air alongs with it, even as, after a showr of Rain, a spait of Water comes, and carries alongs with it, both the foul Water and the clean, it meets with. Though this may be probable, which seems to be your own opinion, yet the other seems to be more probable.

The other sort of *Damp*, is that which they call *want of Air* ; and though the term be not altogether proper (there being no space without some Air) yet there is a want of Air, which is sufficient for respiration of *Animals*, or for the burning of fire. This is ordinarily found in the running of *Mines* under ground, for conveying of Water from *Coal*, or other *Metals*, or in the *waste Rooms* of *Coals*, where the *Sinks* are very deep, and to evite the charge thereof, there is ſome neceſſity to work as far under ground for winning of *Coal*, as is poſſible, without new *Sinks*. The cauſe ſeems to be, that the Air under ground, in ſuch caſes, wants communication with the Air above ground, becauſe it is found, that by giving more communication, the evil is cured. Whence comes the neceſſity of *Air-holes* in *Levels*, which are ſo many *Sinks* ſet down, for no other uſe, but for giving Air to the *Workers*. Some are of opinion, that this defect might be ſupplied by the blowing of Bellows, from above ground, through a Stroop of Leather, or of ſome other thing, which muſt run along to the end of the *Level*, for keeping the Air there in motion. But I have not yet heard, that it hath been made practicable.

The effects of this *Damp* are not ſo dangerous, as theſe of the other. 'Tis true, it will kill Animals, and extinguiſh burning *Coals* and *Candles*, but not ſo ſuddenly as the former; and ſo people are not ſo readily ſurprized by it. The other ſeems to kill by ſome poiſonous quality : in this *Animals* dies for want of ſufficient Air for reſpiration. Therefore in advancing in a *Coal Room*, or *Level* where this is, you ſhall ſee the flame of the Candle grow leſs and leſs by degrees, till at laſt it be totally extinguiſhed, and the perſon entering, ſhall find the difficulty of breathing grow greater, as he advanceth forward, till at laſt he cannot breath at all. Hence it is, that few or none are killed by this kind of *Damp*, and all its prejudice is, that it renders the work more chargeable, when there is a neceſſity to remove it.

For that, which they call *Wild-fire*, it being a thing not incident, but to very few *Coals*, is leſs known, than any of the reſt of the accidents that follows *Coals*. The ac count I have heard of it, is, that in some *Coals*, which naturally are full of *Oil*, and that are (as they call them) *fatt Coals*, there is a certain *Fire*, which is as a *Meteor*, and I judge, that from its resemblance to *Ignis fatuus*, which the Vulgar termeth *Wild-fire*, it hath the same name. It seems to be composed of some *fatt oily vapour*, that goeth out of the *Coal*, the *Pores* thereof being once opened, which is kindled after the same manner, as those fires above ground are, which are most ordinarily found in fatt, and marrish ground. Of this fire it is reported, that in the day time, while the *Work men*, are working in the *Coal-roomes*, it comes to no height, though it be sometimes seen in little holes of the *Coal-wall*, shining like kindled sulphure, but without force: but when the *Work-men* are once removed, and have stayed out all night, it gathers to such a strength, that at its first encountering with fire, which the *Coal-hewers* are necessitate to have, by taking in of light, it breaks out with such a violence, that it kills any person, it finds in its way. The reason, why it is without this force, while the *Work-men* are in the place, seems to be this, that they working with such violence, and motion as they do, do certainly move the Air considerably, it being contained in so narrow a place, as a *Coal-room*. And this Air being violented by motion, moves that *oily vapour*, whereof the fire is formed, so that it gets not liberty to unit it self, being dissipated by the motion of the Air. But so soon, as the Air is still, and quiet, after the *Work-men* are gone home, it units it self, and gathers force, and therefore, so soon, as it meets with fire, which is more forcible, than the flame that is kindled in it, it rarifieth; the sulphurious parts being kindled, and forceth it self out, as powder out of a *Gun*. For it hath been observed, that if any person stay in the *Coal-*

sink while it breaks within the *Coal-room*, they are in danger of being killed. The ordinary way by which the hurt of it is prevented, is by a perſon that enters, before the *Work-men*, who being covered with *wet ſack-cloath*, when he comes near the *Coal-wall*, where the *Fire* is feared, he creepeth on his belly, with a long *Poll* before him, with a lighted candle on the end thereof, with whoſe flame the *Wild-fire* meeting, breaketh with violence, and running alongs the *roof*, goeth out with a noiſe, at the mouth of the *Sink*, the perſon that gave fire, having eſcaped, by creeping on the ground, and keeping his face cloſe to it, till it be over-paſſed, which is in a moment.

The place, where this was moſt known, was in a *Coal* be-weſt *Leith;* in a piece of Land called *Werdy*, which for want of *Level*, and the violence of that *Fire*, the *Owners* were forced to abandon.

I come now to the laſt part, which I promiſed to ſpeak of, namely of the beſt way for trying of grounds, to find *Coal*, where never any hath hitherto been diſcovered, and of carrying on of *Levels*, for draining the Water of *Coals* and making it workable. As to the firſt part, there are but three wayes. Firſt by *ſinking*, which is moſt chargeable, in regard, that in ſuch grounds, where the *Metals* are all intire, Water abounds, and this doth not only bring the *Maſter* under a neceſſity of great expence for drawing the Water, but alſo rendereth it impoſſible to get *ſinked* to any deepneſs, which may ſuffice, for giving an account of all the *Metals* to be found, within the field, that may be rendred workable. There was a ſecond way invented to ſupply this defect, which is by *boaring*, with an inſtrument made of ſeveral *Rods* of Iron, which boareth thorow the *Metals*, and tryes them. This way in my opinion, is

worſe

worfe then the former. For firft, if the *Coal* ly deep, in
the place where you try by boaring, it becomes almoft as
tedious, and expenfive, as *finking*, the drawing of the
Rodes, confuming fo much time, in regard it muft be frequently done. Next, in *boaring*, fuppofe the nature of
the *Metals*, be found, yet thereby their *courfe* can never
be known, till they be *finked*, which is one of the things
moft confiderable in the fearch of a *Coal*, becaufe thereby
is known, whether it be workable, with advantage or not,
and whether it be poffible to draw Water from it by a *Level*,
or otherwife. Laftly, this way leaves the *Mafter* at an
uncertainty (notwithftanding the *Coal* had been found)
of its *goodnefs*, as to its *nature*, and as to its *thicknefs*. As
to its *goodnefs*, becaufe all that is found of the *Coal*, by
this *boaring inftrument*, is fome fmall *drofs*, which remains
after the wafhing of the thing that's brought up in the
wumble, by which none can judge of its *goodnefs*, or *bad-
nefs*. As to its *thicknefs*, becaufe it is impoffible to difcern
exactly, when the *boaring-inftrument* hath paffed the *Coal*:
all the rule for trying thereof, being the kind of *Metal* that
is brought up in the *wumble*. Now, I have known in my
experience a *Coal boared*, which the *Boarer* by that rule
hath judged four foot in *thicknefs*, yet when it came to be
finked, hath not proven one. The reafon whereof, is obvious, becaufe the *boaring-irons*, being long, and weighty in lifting them up, and down, they break the *Coal*,
already pierced ; and this falling down among the *Metals*,
they are piercing, and being found in the *wumble* with
them (efpecially when the *Metal* under the *Coal*, is a *black
Till*) gives ground to imagine, that all that time, they
have been peircing a *Coal*, and fo confequently, the *Coal*
muft be of fuch a *thicknefs*.

The

The laſt, and beſt way of trial, is that which is termed an *ranging over the Metals.* For doing whereof, this method, is to be obſerved. Suppoſe there be any place within the ground to be ſearched, where the courſe of *Metals* can be ſeen, as in the *banks* of a *River,* or *Rivolet,* or *Seabanks,* when the place is near the Sea, then conſideration muſt be had how far the loweſt of theſe *Metals,* can go before they *Crop* out to the *Graſs,* which will be known by obſerving the *Dipp* or declination of the *Metals,* and the *Riſe* of the ground above, whereof a juſt allowance muſt be given, and having *digged* before the ſaid *Crop,* you ſhall certainly find, the *Metal,* that is next under it, and if that prove not *Coal,* keeping the former proportion, you muſt advance, and *digg* before its *Crop,* and ſo ſhall you find, the next *Metal* under it, and ſo ſtill, till you have tried your ground, and found the *Crops* of all your *Metals* within it. But if there be no *Water-banks,* or ſuch like, to give you the firſt view, of the courſe of your *Metals,* then muſt you *ſink* firſt at *random,* and having once paſt the *Clay,* you will readily overtake ſome *Metals,* whereby you will know the *courſe* of the reſt, and having once found the *Dipp* and *Riſe,* you muſt follow the method of *ranging* already preſcribed, except the ground ſo to be tried, contains not within it ſelf the *Crops* of the *Metals,* the body whereof lies in it, whether of *Coal,* or *Stone,* in that caſe, there is no way to try, but by *ſinking,* or *boaring.* The way of *ranging* is conſpicuous in the following figure.

Figure

Figure 11.

P   O   K   L   M   N

F   G   H   I

C   D   E

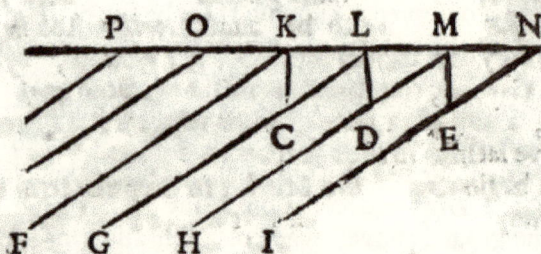

The piece of ground to be tried, is P N, where there
are several *Seams* of *Metals*, that *Cropps* out at the Points
K L M N. Suppose the lowest to be the *Coal, viz.* I N,
for which you are to make trial. You *Digg* first at K,
without the *Cropp* of the *Seam* F K, and you *dig* till you
find the other *Seam* of *Stone* G L, at the Point C. Following the Rule before given, you advance before its *cropp*,
and *diggs* at L, and finds the other *Seam* of *Stone* H M, at
the point D: from which you also advance, and *diggs* before its *cropp*, at the point M, and finds your *Coal* at the
point E. But, if by advancing over the *cropps* of these
Metals, which comes out from under one another, you
find no *Coal*; then you are to *range* backward, for the
cropps of *Metals* lying above these, where haply the *Coal*
may be, as at O, and P. This in my opinion, is the most
certain and exact way of trying Fields for *Coal*, or any other
Metal of that nature, and least chargeable of all others.

The second of this last part, I promised to speak of,
was in order to *Levels*, or *Coal-Mines*, which are nothing
else, but *Conduits* or *Gutters* made under ground, for conveying of the Water from the *Coal*, and so rendering it
P p   work-

workable. It seems that a very little time before this, that way of *Mineing* under ground hath not been fallen upon. For there are to be found *Coals* wasted in their Cropps only; for conveying the Water whereof, they have made a *Conduit*, or *Level*, which hath been *open to the Surface*, like a great *Ditch*, some whereof have been ten or twelve fathom in their deepness.

The beginning of the *Level* (to keep the term used) must alwayes be at the lowest part of the Field, where the *Coal* lyes to be dryed. Some whereof, by the *rising* of the ground, and the *Streek* of the *Coal* rising that way (as we shew before) gives the advantage of a *Free Level*, that is, when the Water comes above ground of its own accord, without being forced by drawing. In others, there is a necessity of *Engines* to draw the Water from the lowest part of the *Level*, and bring it above ground; which *Engines* are of several sorts. As when *men draw* with ordinary *Buckets*, or when there is a *horse-work*, or *water-work*, and that either by a *Chain* with *Plates*, and a *Pump*, or with a *Chain* and *Buckets*; all which are very common, especially those we have in *Scotland*, they being capable to draw but a very small draught, making only use of one *Sink* for that effect. But there are to be seen in the North of *England*, in *Bishoprick*, *Water-works*, by which Water is drawn above 40 fathom in perpendicular, but not all in one *Sink*. The manner whereof is thus, there being a *Sink* from the end of their *Level*, to the surface of the earth, where their *Works* are going, 40 fathom deep, which must dry the *Coal-Sinks* at 60 or 70, which ly above the *Banks* of the *River*, where the *Water-works* are scituated, there is first one 40 fathom deep from the *Grass*. Another in a right Line from that, of 24. Another of 12; upon all which
there

there are *Water-works.* In the first *Sink* the Water is drawn
from the bottom 12 fathom, and thence conveyed into a
Level or *Mine,* which carries it away to the second *Sink.*
By the second *Work,* the Water is drawn out of the second
Sink 14 fathom, from the bottom, and set in by a *Level*
to the third *Sink,* which being only 12 fathom deep, the
Water-work sets it above ground. The form of the *Engine*
is after this manner. In the first *Sink* there is an *Outter-
wheel* moved, as other *Milns* are, by the Water of the
River: upon the end of the *Axle-tree* of which *Wheel,* there
is a *Ragg-wheel,* turning *vertically,* as doth the *Outer-wheel.*
This *Ragg-wheel* by a *Nutt,* or *Trinle* turns another, which
moves *horizontally,* the *Axle-tree* whereof goes right down
in the *Sink,* and may be is 8 or 10 fathom; at the end
whereof there is another *Ragg,* which by a *Nutt* turns another *Wheel,* which goes *vertically* as the first *Ragg,* and
causeth another Wheel with a long *Axle-tree* turn as the
first, and so down till it come to the *Wheel,* which turns the
Axle-tree, by which the *Chain* is drawn. The second
Sink, hath such another *Engine,* but not so many *Wheels,*
in regard it is not so deep. The third, hath only one *single
Wheel,* whereby the Water is drawn above ground.

The most curious of these *Engines,* that are to be seen,
are at *Ravensworth* near to *Newcastle,* which belongs to
Sir *Thomas Liddel,* a most ingenious Gentleman, who, for
procuring a *Fall* of Water, which may serve the *Wheels* of
all the three *Sinks,* hath erected the first work upon Pillars like a *Wind-Mill,* pretty high above ground, from
which the Water falling, makes the second go close above
ground. And to make the Water fall to the third, the
whole *Wheel* is made go within the *surface* of the ground,
which *terminats* at a *River* under the *Works,* which *Mine*

is

is of a confiderable length. Where Water cannot be had
to make such *Works* go, they use *Horse-works*, but not with
fo good succefs, being more chargeable, and not having
fo much force and power, as the *Water-works*. But I am
of opinion, that *Wind-works* might ferve well, where Water cannot be had; and when no Wind fhould happen to
blow, the fame *Works* might be fupplied by *Horfe*: and
that the *Wind*, when it blows but ordinarily, hath as much
force, as fo much Water, which is made ufe of for turning
fuch *Wheels*, is to me unquestionable. For I have feen in
Holland, a *Wind-Mill*, that by the motion of the *Outter-
wheel*, caused feven pair of *Mill-ftones* to go at once, befides another motion for bringing the *Victual* from the
ground, four or five Stories high, to be *Grund*. And feveral *Saw-Mills*, which befides fix or feven great *Saws*,
they caufed go, did by another motion bring up from the
Water great *Trees* like *Ship-Mafts*, to be *fawen*, and placed
them right againft the *Saw*; all which could not be but of
greater weight, than 10 or 12 fathom of *Chain* with *Buckets*, or *Plates* for drawing of Water.

But to return, for the right making of a *Level*, the true
hight of the ground, where the *Coal* lyes muft be firft taken, that it may be known, how much of the field can be
drained by it; which muft be done, either with a *Quadrant*,
or with an Inftrument made exprefs. Then care muft be
taken, to take the loweft part for the mouth of the *Level*,
that the field can afford, and from that it muft be carried
in a ftraight line towards that part of the field, where the
Coal is thought to be encountered by the *Mine*. In working whereof, two things are in a fpecial manner to be re-
guarded. Firft, that the *Level* be wrought without *afcent*,
or *defcent*: the beft way for trying this, being by the furface

face of the Water paffing through it, which ought to be as
little moving, as can be : for the lofs of one foot of *Level*,
which the ground gives, is a lofs of a confiderab'e parcel
of *Coal* to be digged, efpecially if it be *flate*. If there occur
any *Metals*, which are impregnable, in the *courfe* of the
Level, fo that it is impoffible, to follow fo ftraight a line, in
regard the *Mine* muft be wrought over the top of that
ftone, which is unworkable, in that cafe, there is but one
of two to ferve the lofs of *Level*; either the *Coal* rifes in
Streek towards which the *Mine* is carried, and if that be,
then after that *ftone* is paft, the *Level* muft be carried, as
low, as it was before it encountered the fame, and the
courfe of the Water fhall not be obftructed, becaufe the
fourfe, viz. the *Coal* from whence the Water comes, rifing
higher than the *Stone*, the Water fhall eafily pafs over that
hight. Hence it is, that we fee in fome *Coals*, that have
been wrought, at the loweft point of their *Streek* by a
drawing-fink, and the *Streek* rifing from that point, the
Water that hath come off the *Coal*, being in its *Sourfe*
higher, than the mouth of that *drawing-fink*, hath made
it to over-run, and ferve to difcharge all the Water, that
comes therefrom. But, if the *Mine* be run to a *Coal*, that
after it hath overtaken it, rifes no higher in *Streek*, than
the *Mine* it felf, the Water that comes from it, will not
pafs over any hight in its way, but will be unqueftionably
ftopped. Therefore, in cafe fuch an impediment could
not be removed, as many times fuch *Metals* will fall in,
which are unworkable in a direct line, the ufe of a *Siphon*
might be tried, which would unqueftionably fupply the
lofs of about 32 foot of *Level*, this being the hight in Perpendicular, to which the *Preffure* of the *Air*, is able to raife
Water up thorow a *Siphon*.

The

The next thing to be observed in carrying on of *Levels*, are the *Air-holes*, for which there is a necessity indispensable. In setting down whereof, care must be had, that they be not directly upon the *Mine*, lest *rubbish* falling thorow from above ground, should stop, and obstruct the same, and so obstruct the *course* of the Water; and therefore it's better they be set down at a *side*, their only use being to communicate fresh Air to the *Work-men*, which if it could be otherwise supplied (as I think it not utterly impossible) would render the charge of the *Coal-works* a great deal more easy.

Other things might be spoken to of *Levels*, as that some run with the *course* of *Metals*, they pass thorow; and that some run against that *course*; and of bringing *Level* from the *Dip* of an *upper-Coal*, which hath a *Level* of its own, to dry a *Coal* lying under it, which cannot be otherwise done. But these things being common and obvious to any, who have but the smallest skill and experience, I shall forbear.

This confused account, your importunity hath drawn from me, for which if your *Book* suffer censure, which I grant it may do, as to this part of it, you are to blame yourself, and so I rest and am, *&c.*

FINIS.

POSTSCRIPT.

Reader,

THat thou mayeſt know the riſe, and occaſion of this Poſt-ſcript, which I have ſubjoyned, I ſhall give thee this ſhort account. When this Book was fiift committed to the Preſs, I ſent an intimation thereof to ſeveral perſons, whom I judg-ed would encourage it, yet to none, but to ſuch, in whoſe kind-neſs I had confidence, and whom I judged my real friends . Among others, I ſent over to Saint *Andrews* one of my Edicts, to one or two there, in whom I truſted, but in ſtead of a kindly return from them, to whom I had written moſt affectionatly, they wrot back a Letter, wherein they ſuperciliouſly condemn the purpoſes of this Book, before ever they had ſeen them, which is as follows.

Sir,

I Received yours on Saturday *laſt, and having occaſion the ſame night to be in company with many of the* Maſters *of the Univerſity, I made known your reſolution to them, ſhewing them your Edict, and de-ſiring their Contributions: ſome were not pleaſed, that ye call the Do-ctrine concerning the weight and preſſure of the Water in its own Ele-ment, new, ſeing* Archimedes *hath affirmed, and demonſtrated in his Books* de inſidentibus humido *the ſame Geometrically* 2000 *years ago ; others affirmed that it was ſo far from being new, that they would undertake to demonſtrat the event of any of all your Experiments* à priore *from* Archimedes *his grounds, yea, in general of any Hydro-ſtatical Experiment, ſeing they look upon it,* as a Science long ago per-fected. *Some ſaid, as to* Diving, *that they imagined any me-thod better then that of* Melgims, *which is now vulgar, to be impoſ-ſible. As to the Obſervation of the* Sun, *or Moons motion in a ſecond*

Q q

of

of time, yea, or much less, it can be done most exactly by a Telescope, and a Pendulum, but serves to no purpose, seing that same motion can be had infinitly more exact by proportion, from observations of a considerable interval ; for so the Astronomers collect all the middle motions of the Planets. As for the Observations of Coal-sinks, latitude of Edinburgh, *and its variation of the Needle, they may assuredly increase the Historical part of Learning : yet many of the Masters here imagine themselves concerned in credit not to promote the publication of any thing, which seemeth to declare our Nation ignorant (by calling them new, and unheard of) of these things known over all the World these many years among really Learned Men, albeit they be debated amongst ridiculous Monkish Philosophers. I conceive, ye would do best to undeceive this University, by sending us some of your most abstruse Theorems, and surprizing Experiments; which if they be not evidently and clearly deduceable from* Archimedes, *or* Stevinus, *who did write long ago, or rather, if they be not the same with theirs : ye may assure your self that this University will take away at least all the obligations ye have sent here ; otherwayes, I am afraid, I shall not be able to prevail with them. I hope ye will pardon this my freedom I use with you, and return an answer with the first occasion, to*

   St. *Andrews, De-*   Sir,
     cemb. 27. 1671.   Your most humble Servant.

After the receit of this, being unwilling to make it a ground of debate, I returned a most discreet answer, thinking to conquer their humour with civility, and kindness, but not long after, hearing of their clamour against the Intimation , and of their disswading others, who would willingly (I suppose) have condescended, I was necessitated to send this return, for a joynt answer to them both, for besides this, another of the same kind came also, of which hereafter.

                                                     Sir,

Sir,

I *Received yours, of the Date of* December 27. 1671, *and though it was a little unpleasant, yet I took it very kindly from you, as from a person I judged ingenuous, as my return of* January 9. 1672. *can witness, wherein I did not in the least resent any thing you wrot ; neither would I ever have done, if you, and some others especially with you, had not proclaimed publickly, what you and they had written to me privatly, the noise whereof, I have heard here, by several persons who came from the place. Therefore, Sir, you must pardon me, if now at last, after so much silence, I return you this answer, for no other end, but for my own vindication, in what I have lately Printed, and am about to Print. I am very much then surprized with the answer, that you and they have returned, such a rank smell of prejudice and envy, I find in it. I am rewarded evil for good ; for I minded nothing but good-will ; else, you and they should never have been troubled with my proposal. If they had affected the reputation of Learning, there was another way to it, then the course they have taken, namely to condemn with such a deal of superciliousness, as derogatory to the credit of the Nation, forsooth, the labours of one, that hath done more for the credit thereof, then they have done as yet. They might have minded the saying of the grave Historian,* Nam famam atque gloriam, Bonus atque ignavus æque sibi exoptant: ille verâ viâ nititur, huic, quia Bonæ artes desunt, dolis atque fallaciis contendit. *And for undeceiving of the University, as I am very far from counting such persons the University, so have I more respect for it, and all Learned Persons in it, then to account their deed, the deed of the University. As for what they can do, for promoting the work I have now at the Press, I value it not at the rate of shewing them so much as one of my Theorems : for, if they have snarled so much, but at one word, in the intimation of the work ; what would they do, if they had more of it ? which yet must stand firm, unless they (for 'tis a matter of fact, and cannot be contradicted with* Sophistry *and* Non-sense*) overthrow it, which I little fear, as* Cicero *did* Verres, Tabulis & Testibus ad singula indicia prolatis. *Neither will their imagination do it, for that cannot make* factum infe-

ctum

&um. *It seemeth to be a great weight, that they lay upon the force of their imagination, since they are so confident, as to say, they imagine any method of* Diving *better then that of* Melgims, *to be impossible,* adeo familiare est hominibus supra vires humanas credere, quicquid supra illorum captum sit. *As for these others, that would demonstrat à* priori, *the event of all my Experiments from the grounds of* Archimedes, *as I doubt not, but they would, if they could, so in this they bewray their want of skill: for* Archimedes *wanted a necessary requisite, which I go upon for my deductions. And though it were true (which they say) that all my* Theorems *were demonstrable à* priori *from the grounds of* Archimedes, *yet this doth not hinder them to be both new, and un-heard-of, as if new, and un-heard-of conclusions, might not be deduced from old principles. In this they are so much the better, and not the worse. And whereas they say, they look upon the* Hydrostaticks, *as a Science long since perfected, in this they do yet more discover their weakness: for what one Science hath yet come to its perfection? Nay, hath not this* Pedantick *humour been the great bane of good Learning, that Sciences were already perfected? So that* Seneca *said truly,* Puto multos pervenire potuisse ad sapientiam, nisi putassent se pervenisse. *As for the representing of the* Sun *or* Moons *motion to the eye (for that should surely have been taken in) that you say, serveth to no purpose, to me is a little uncouth, considering how much it conduceth to the accuracy of Astronomical Observations, beyond what the former Ages could attain to. And whereas you say, it can be had infinitly more exactly by Observations of a considerable interval, as Astronomers collect all the middle motions of the* Planets, *but I say, even those intervals should have been far better known, if they had by this mean, and the* Oscillatory Clock *been observed; so whatever arguing by the rule of proportion, may do for shewing the* Suns *motion in* seconds, *and* thirds, *it reacheth not these accuracies, that are reached by this invention, so long as the* Sense *cannot deprehend, and fix them. As for the Observations of* Coal-sinks, *&c. which you say, may assuredly increase the* Historical *part of Learning; are they not for this the more useful, since the Scientifical part of Learning dependeth so much on the* Historical *part, and which conduceth more thereto, then all the pre-*

carious

carious principles of Cartesius, Epicurus, *and the like* ; *who in stead of giving us an account of the World that* God *made* , *have given us imaginary ones of their own making :* so *that such a History*, *as* Natural Philosophy *requires*, *is wisely accounted among the* desiderata *in Learning by all sound Philosophers to this day*. *So much in answer to yours*, *and I rest*

Edinburgh, Feb. 22.
   1 6 7 2.
                           Your Servant,

IN answer to this last, there came to my hands from *St. Andrews* a Letter unsubscribed by any Master, full of barbarous railings, passing all bounds of civility, against my self, friends, and works, which, if the *Contrivers* had not been more gall'd with reason, then injuries, I suppose they would have forborn. And thinking this not sufficient, they would needs aggravate the wrong, by one circumstance more, which they either did out of *disdain*, or *fear*, not daring to own what they had contrived, in making the *Bedale* of the University subscribe it. And to give a further proof of their insatiable malice, they must needs distribute copies thereof, as glorying in their shame, one whereof was sent over to *Edinburgh* unsubscribed also. Now, let any indifferent person judge, whether or not , I have not reason to do what I have done. They have been the first proclaimers, though in a *clandestine* way, and why not I next, in this way. But left, they think, they have marred as much the tranquillity of my mind therewith, as they have their own, I shall answer in the words of the Moralist, *Eleganter* Demetrius *naster solet dicere*, *eodem loco sibi esse voces imperitorum*, *quo ventre redditos crepitus*. *Quid enim inquit*, *mea refert*, *sursum isti*, *sive deorsum sonent*. And let this stand, for the railing part of the letter.

But first, whereas he should have spoken to the contents of this Book, he falleth foul upon my last Peice, intituled, *Ars nova*, & *magna*, *gravitatis*, & *levitatis*, snarling eight or nine times, at the bare title, like a *Cur* at the horse heels, when he cannot reach the rider. This lay not in his way, doing herein like *Vejento* the blind

Courtier

Courtier of *Domitian*, who, when he should have turned his face to
the right hand, where the *Sturgeon* lay, turned it to the left.

      ——————— *Nam plurima dixit*
     In lævam conversus : at illi dextra jacebat
     Bellua.

So that concerning all these *invectives*, I may say, *sed quid hæc
ad Rhombum.* But what other can be expected, *ubi furor arma mi-
nistrat.* But seing his Letter shews, how sick he is of the plague
of malice, and envy, I am so far from storming at him, that I pity
him, though he may be a Master, and teacher of others, and wish
him to teach himself. ——————*Servitium acre*

   Te nihil impellit ? nec quicquam extrinsecus intrat
   Quod nervos agitet ? Sed si intus & jecore ægro
   Nascantur domini, qui tu impunitior exis
   Atque hic, quem ad strigiles scutica & metus egit herilis.

That I do not interpret this (Reader) excuse me, for I am
speaking (I suppose) to a Master of an University, and a gentleman too, of very high pretences, as to learning. And yet I cannot but think strange of two things. First, that he returneth not
the least *Latine* sentence in answer to mine, no not so much as pertinent language in his *Mother-tongue.* What? An University-
man, and no return in *Latine* to these sayings, of so grave Authors,
or at least in pertinent English. The other, that he no more un-
derstands, these words, as *Cicero did Verres, tabulis & testibus ad
singula indicia prolatis,* than the *Curat* did the *Modicum bonum*
that he was desired to prepare for the Bishops dinner. For, whereas
he saith, *as for your Latine sentences, where ar our doli, and fallaciæ,
tabulæ & testes, sapientia ad quam putamus nos perveniße.* To pass
the first and last question, of which anone, the second was most
improper for him to ask at me, who did put him to it, to overthrow the title of my Experiments, to wit *New*, not by *Sophistry,* and *Non-sense,* but as *Cicero* did *Verres, tabulis & testibus,*
by proof and Witnesses; this he should not have asked, but answered. I am confident a Boy in the *second* Class, could better
have understood these words, than this man. And for the first
question, *where are our* doli, *and* fallaciæ ? Why should he ask it,
                                        seing

seing the design of his Letter may be evidently seen, to put *Royal Societies*, and *Universities* between him and me, in the front, whom I have not made my party, but to whom I owe all due respect, and such a poor pitiful fellow as the *Bedale* in the *Rear*, in causing him subscribe his letter thus,

March 14. 1672.   *Mr.* Patrick Mathers, *Arch-bedale to the University of* St. Andrews.

Is not this to do, as the *Butcher* did, who sought his knife, when it was sticking in his teeth. If the University ordered this subscription, it would have been said, *at the command of the University*. If not, it cannot be purged from a false insinuation: and the University may justly resent it, that their publick servant, hath been so abused. If the fear of a counterblow hath made him afraid, to put his hand to it, he hath done as the *Ape* did, that thrust the *Cats* foot into the fire, because he durst not do it himself, and given a palpable discovery of the diffidence he had of his cause. If he hath done it, to put indignity on his adversary, he hath missed his mark, for as a certain Writer saith well, *Infamy is as it is received*. *If thou be Mud-wall it will stick: if Marble it will rebound: if thou storm at it, it is thine: if thou despise it* (as I do this) *it is his*. But besides this, he endeavoureth to put Mr. *James Gregory* between him and me also, and bringeth him in speaking of my writings, with such a deal of disdain and sauciness, *ut nihil supra*. What? was Mr. *James Gregory* such an eminent person, that he could not speak his thoughts himself, but needeth you Sir, for a *Proxy*, and Chancellour to speak for him. If Mr. *James Gregory* will speak to me, what you have spoken in his name, he shall have an answer. But I have no mind to gratify so far your *doli*, and *fallaciæ*, as to fall on any man upon your word, having so little confidence of your common honesty. This were *perversam gratiam gratificari*. Wherefore passing his impertinent railings, I come to answer, what he hath returned to my purposes in my last. And that he may get no wrong, I shall set down the very words of his Letter, *viz. as to what you write concerning the imperfections of Sciences: the Scientifical part of Geography is so perfected, that there is nothing re-*

quired

quired for the projection, description, and situation of a place, which cannot be done, and demonstrat. The truth is they have overshot themselves in this, though they be ashamed to acknowledge so much; for what a pitiful shift is it, to bring *Geography* for an instance of a perfected Science, when so much of the Earth remains to this day unknown altogether, as the *Universal Mapps* testify. Of the known parts, how little is there to this day sufficiently described by the exactest *Mapps*, that time, and labours of men have yet produced. And now to retort your own question upon your self, *ubi est sapientia ad quam putatis vos pervenisse?* O but saith the Author, *it is perfected as to its scientifical part.* But I pray you *Sir*, what is this, though you may be a teacher of *Logick* of no small esteem with your self, and disdain of others, but to play the *Sophister*, by the *Fallacy*, *à dicto secundum quid, ad dictum simpliciter*: *Geography is perfected as to its scientifical part, therefore it may be called a perfected Science*, when it is so defective as to the *Historical part.* If *Astronomy* to this day be a Science not perfected, through want of its *Historical part*, shall not *Geography* be so likewise. But furder *Sir*, for the *Scientifical part* of *Geography*, which you alledge to be perfected, in this also you argue against the rules of *Logick*, in committing that same *Fallacy* over again, for giving and not granting what you say, that the *Scientifical part of Geography* were perfected, as to the *projection, description, and situation* of a place, is it for this perfected as to the *Scientifical part simpliciter*, which you are obliged to prove, else you say nothing to the purpose. And what I pray you, is that poor alleadgence you make, in comparison of these things, wherein *Geography* is defective, even as to the *Scientifical part?* Who hath spoken yet sufficiently to the *surface*, and hight of the *Sea* above the *Earth*, the hight of the *Hills*, and *Mountains*, *Longitude* of places, nay the circumference of the Earth it self? Answer this question, if you can, *Hast thou perceived the breadth of the earth, declare if thou knowest it all?* Job. 38. 18. And now *Sir*, I must put you to it again, *ubi est sapientia ad quam putatis vos pervenisse.*

His next answer runneth thus, *The Scientifical part of the Opticks is so perfected, that nothing can be required for the perfection of the sight,*

sight, which is not demonstrat, albeit mens hands cannot reach it.
And these being the objects, of the foresaid Sciences (you should *Sir,*
have said, the whole objects of the foresaid Sciences, else you still
play the Sophister) *your authority shall not perswade him, or us, that*
it is altogether improper to call them perfect. But mark Reader,
how the force of reason maketh these *Authors* to succumb : for
whereas they should have said, *that it is not improper to call them*
perfect, they qualify it thus, it is not *altogether* improper. And
again, *your authority shall not perswade us , that it is altogether im-*
proper. But (*my Masters*) I do not crave that my authority
may perswade you, but reason. Wherefore to return : *the Scien-*
tifical part of the Opticks (say they) *is so perfected, that nothing*
can be required for the perfection of the sight, which is not demonstrat,
albeit mens hands cannot reach it. But where *Sir,* and by what
person is this done? Shew me the man, (if you can) that hath done
it. But though all this were true, were therefore, either the *Op-*
ticks, Dioptricks, or Catoptricks perfected Sciences? Who hath
yet sufficiently explained the manner *how we see,* far less how *Birds,*
and *Fishes, Beasts,* and *Insects* see? How the *Eagle* mounting aloft
spyeth her prey from a far. Who hath spoken sufficiently to the
nature of *colours?* For these also belong to the *Opticks,* or of *light,*
and of the *infraction,* and *refraction* thereof. The learned *Lord*
Verulam was not of your mind *Sir,* when he wrot thus, *De forma*
lucis, quod non debita non facta fuerit inquisitio (præsertim cum in
Perspectivâ strenuè elaborûnt homines) stupenda quædam negligen-
tia censeri possit. Etenim, nec in perspectivâ, nec aliàs, aliquid de
luce, quod valeat, inquisitum est.

If Mr. *Newton* had been of this *Authors* mind, he should not have
attempted the late invention of his *Span-long Dioptrical-catoptrical*
Prospect, whereby *Jupiter* his *Satellites,* and *Venus horned* are to
be seen. And if Mr. *Hook,* had been of his mind, he should not
have made his late *Proposal* of *Telescopes, Microscopes, Scotoscopes,*
by figures as easily made, as those that are plain and spherical,
whereby the *light,* and *Magnitude* of *Objects,* may be prodigiously
increased, and whatsoever else hath hitherto been attempted, or
almost desired in *Dioptricks,* may be accomplished. Where observe

ferve (Reader) how that ingenuous perfon, is fo far from the *windy language* of this Author, that he doth not fay, *whatfoever can be required for the perfection of fight is demonftrat*, or any thing like it, but *whatfoever hath been hitherto attempted, or almoft defired.* For who can tell, what fhall be found out hereafter, even in thefe things. To them we may borrow the words of the Poet,

Prudens futuri temporis exitum
Caliginofa nocte premit Deus.

So, Sir, I ftill put you to that queftion, *Ubi eft fapientia ad quam putatis vos perveniffe ?*

In the next place he falleth upon the *Hydroftaticks*, which formerly he looked upon as a Science perfected long ago. But becaufe in his anfwer, he in effect yeelds the caufe, I purfue him no further. *Habemus confitentem reum*, while he exprefly grants, *there are many things yet* (faith he) *relating to the proportion and acceleration of the motion of Fluids , which are yet unknown.* As for his reflections upon what I have written in my *Ars Nova*, concerning a perpetual motion, which I never intended to demonftrat, I leave them as *indicia ægri & impotentis animi.* I proceed to anfwer him in what he addeth thus. *Only we cannot but admire your fimplicity in this, Aftronomy feeketh alwayes to have the greateft intervalls betwix obfervations, and ye take that ye will give an excellent way for obferving the Sun or Moons motion for a fecond of time, that is to fay, as if it wer a great matter, that there is but a fecond of tym betwix your obfervations. I wonder yow fay the eye fhuld be added, for the invention had been much greater had that been away.* But what is this Sir, but ftill to play the *Sophifter ?* Is not this the *Sophifm, ab ignoratione Elenchi ?* for it doth not contradict my conclufion, which is, that Aftronomical Obfervations, by this mean, and the Ofcillatory Clock, may be made to a fecond of time, which is of fo great importance in *Aftronomy.* But mark the *Non-fenfe* (Reader) *the invention* (faith he) *had been much greater, if the eye had been away*: that is, the invention of this Obfervation had been much greater, if the *eye*, that is, the *Obfervation* had been away. In this they have outfhot themfelves alfo ; and what they fpoke
unadvifedly

unadvisedly before, they will now speak deliberatly, and defend it rather by *Sophistry* and *Non-sense*, then yeeld to the truth.

Has toties optata exegit gloria pœnas.

The Author addeth, *None will denay but that an guid history of nature is absolutelie the most necessary requisite thing for learning, yet it is not like, that yow are fit for that purpose, who so fermelie beleeves the myrakles of the Vest, as to put them in Prent, and recordeth the semple Meridian Altitudes of Comets, and that only to halfs of degrees, or little maire, as worthy noticing.* If it were needful, I could produce the passages of some of the most Learned Writers, of these last times, that have recorded the like. Were they therefore unfit to write History? A person of this Authors reading and learning, will soon find them out. It he do it not, let him know, that I keep them for a *reserve.* To speak nothing of *Aristotle,* who wrot a Book περὶ θαυμασίων ἀκουσμάτων, extant to this day : was he therefore unfit to write his *Natural Histories?* Prodigious relations, when the memory of them may be found credible, and maintainable, such as mine are, ought not to be excluded from a *Natural History,* or else the Learned Lord *Verulam* is much mistaken in the third *Aphorism* of his *preparatory to Natural and Experimental History.* Nor had he reason to carp at my Observations of the *Comets,* as long as he made none himself. But they will speak for themselves to any that read them. Neither need they him for a *Common Cryer,* either to commend them, or discommend them; who, when I was at these Observations, he possibly hath not been so well exercised.

He subjoyneth, *However if yow do this last part concerning Coalfinks weill, and all the rest be but an* Ars Magna & Nova : *ye may come to gaine the repute of being more fit to be an Collier, than a Schollar.* I must tell this *Pedant,* that a *Coal-hewer* is a more useful person in his own station, to the Countrey, than he is; and that the Science of *Coal,* and other *Minerals,* is far beyond any knowledge this man hath, or can teach: But, my *Lords* and *Gentlemen,* who are *Coal-Masters,* mark this : if ye stand to the judgement of this *Pedant,* though ye had never so much skill in these things, ye may come to gain the repute of being more fit to be

Coal-

Coal-hewers, than *Schollars* ; as if the knowledge of such things were not a part of *Natural Philosophy*. It seems he hath either forgotten the common *definition*, or else hath never known it, that *Physica est Scientia Corporis Naturalis*.

He subjoyneth, *Ye might have let alane the precarious principles, and imaginary Worlds of* Descartes, *till yowr new inventions had made them so : for it man be telled yow* Descartes, *valued the History of Nature, as much as any experimental Philosopher ever did, and perfected it more with judicious Experiments, than ye would do by all appearance in ten ages.* But I pray you, Sir, did *Des-cartes*, and *Epicurus*, and the like, found their *Philosophy* on *Natural History*, and not rather upon their own *precarious principles* : and therefore have quite missed the mark, and method, that was requisite for the advancement of Learning, and have been so far from *grasping* Nature, that it hath *flowen* out from among their hands. As for what he talketh of *Des-cartes*, perfecting *Natural History* by *Experiments*, if he had done it, as the Poet saith in another sense,

Non mihi res, sed me rebus componere conor.

he had done right. But when he took pains on these, to force them to a compliance with his own *fancies*, was not this to study *Natural History*, as *Hereticks* do the *Scripture*, and to be a *Fanatick Philosopher*, and a fit Master for the like of you. The *Proteus of Nature*, must be bound with stronger Chains, then the *fantastick Nugæ of Des-cartes*, before he will tell his secrets. The vanity of whose method may be seen in the *Epicureans*, who having laid down this precarious principle, that *the sense cannot erre*, do turn themselves into so many shapes, to prove *that the Sun is no bigger than a blew Bonnet.*

In end, after he hath given a *Fling* at my labours in *Glasgow Colledge*, about *Universale*, and *Ens rationis*, which I am not afraid he shall come the length of in haste, for ought I can learn, he talleth foul upon the two Lines I cited out of *Juvenal*, in the close of my answer to a passage in a *Philosophical Transaction* : the Lines are,

Cujus

———*Cujus sapientia monstrat*
Summos posse viros, magnaque exempla daturos
Vervecum in patriâ crassoque sub aëre nasci.

Of these Lines, he writeth thus, of which (saith he) *the sense
is not understood, except ye make your self the* summus vir, *and us
all* Verveces. *I suppose this may be the great credit, that ye say, ye
have laboured to gain to your Nation*, viz. *to get us all the honourable
Title of Weathers.* But (Reader) had these *Learned Clerks* been
as skilful in *Rhetorical Composition*, and *Resolution*, as in *Algebraical*, they would not have made such an *Inference:* for the Argument is *à minori ad majus*. Nor was it ever intended for another
end. As for the *honourable Title of Wedders*, which they alledge
I have gained to them, I cannot indeed affirm it ; for if I should,
some surely would judge me to have wronged them as much in this,
as I have done them right all alongs.

But, that thou mayest know (Reader) something more of the
temper of those persons I have to do with in this matter, take but
the following words of one of them, as they are transcribed out of
a Letter written with his own hand to me, after I had written to
him a friendly Letter for obtaining the concurrance of his acquaintance for advancing my Book, *And they promise* (to wit the Masters promise) *ye shall not want their concurrence, whereof ye may be
sure, especially having here your friend* Mr. Gregory, *your Cousin,
and me here to put them in mind. This is all at present, from, Sir,
your real friend and servant.*

Now, what shall be thought of one, who will speak so fair to
your face, and yet cut you with so many invectives behind backs,
let any man judge.

Astutam vapido servat sub pectore vulpem
———*Hic niger est, hunc tu Romane caveto.*

But to give a further discovery of him, in the year 1661, a certain ingenious Gentleman, that had not been bred a Schollar, by
his own industry advanced so far in the *Mathematicks*, that he was
able to set forth an *Almanack*, for which, ingenuous and ingenious
men should have commended him. But this *Author*, with another,
though he had never injured them, and without advertisement, fell

upon him like a couple of *Maſtives*, upon a harmleſs Paſſenger, as if they would have *worried* him in his reputation, in a *Prognoſtica-tion* they ſet forth, rateing and abuſing him out of meaſure : all the cauſe being ſome alledged miſtakes, they thought they found in ſome of his calculations, and in a Table in the end of the *Almanack*, which he calleth *perpetual*, and which they ſay, though falſly, that it will not hold. What had that *righteous man* deſerved at their hands, to be ſo abuſed in Print by them ? But that the deſign is palpable, *the raiſing of reputation to themſelves, upon the ruine of the names of others* ? And yet one of them many years after, was neceſſitat, for fear of *bodily harm*, to crave him pardon, *with humble offer to his knee*. In the *Prognoſtication*, he would needs play the *Poet* in his *Chronology*, which the perſon whom he wronged, might have found more fault with, with better reaſon, than he could do with him, for his Calculations. What a ſtranger he is to the more poliſhed part of Learning, for all his high pretences, theſe Verſes will abundantly teſtify, ſome whereof follow, that thou mayeſt know the reſt, *Tanquam ex ungue leonem*.

Since that the *Julian* period firſt began.
Since that of nought the Lord created man.
He ſhould have ſaid,
Since that of duſt the Lord created man.
He addeth,
Since *Iſrael* from *Egypt* Land did flee.
Since in *Canaan*, he made *Hams* ſons to die.
Since *Romulus* did build his ſtately *Roma*.
Since *Nabonaſſar*, hence is that ancient *æra*.
Since *Gregory* helped the Calendar forlorn, *&c.*

Mark Reader : theſe Verſes are of five feet, at leaſt they ſhould be ſo : but how far he is from obſerving the Precept of that great Maſter of Poets,

Primum ne medio, medium ne diſcrepet imo.

Will appear from his cloſe,
Since fair *Lucina* fulfilled the Golden Number.
Since gliſtering *Phœbus* augmented Sundays Letter.

Euge Poeta.

Ic

It may be he will fay, every man is not born to be a Poet. I anfwer, If the Gentleman, whom he reviled, failed in a calculation, he ought to have been born with, and encouraged: for there are many things that even a mediocrity is commendable in ; but Poefie is none of thefe.

—— Mediocribus effe Poetis
Non Dii, non homines, non conceffere Columnæ.

However, for this, he may affure himfelf, that
Perque Poetarum nunquam celebrabere faftos.

But I leave him to the *Satyrifts* of the time, *Quo illuftrius vapulet,* for his never being feen farder in Print, than by a railing *Almanack,* and ridiculous *Verfes,* the better whereof, might have been made by the Laird of *Dyfert.*

'Tis like this Antagonift, will fet his *Plumbeous Cerebrofity* a work to rifle fome of my Writings, and fhake his head, when he is put to a *demur,* as ever a man did a bottle for *Sack* ; but though he fhould, and I have nothing of his, but an old *Prognoftication* of the Year, 1661, to ripe up, yet who knowes, but I may meet with fome of his *Bajan-notes,* or fome of his wonders about *Ens Rationis,* and *Genus Logicæ,* that he is now fweating at. I am indeed at fome difadvantage, while he only letteth a *flisk* at me, from under deck. Though I have been a little fnell in this reply, yet 'tis no wonder, confidering what a barbarous, and uncivil *Pifle* I met with, which I fhall keep for a referve. I defire to live peaceably with all men. Neither fhall I be foon provocked, fo long as they keep within the bounds of civility. If that be obferved, I fhall thank them, for any miftake they fhall let me fee in my writings, if done with reafon, and without railing.

FINIS.

www.ingramcontent.com/pod-product-compliance
Lightning Source LLC
Chambersburg PA
CBHW021457210326
41599CB00012B/1039